ENERGY POLICIES, POLITICS AND PRICES

U.S. EXPORTS OF OIL AND NATURAL GAS

ANALYSES OF SELECTED ISSUES

ENERGY POLICIES, POLITICS AND PRICES

Additional books in this series can be found on Nova's website under the Series tab.

Additional e-books in this series can be found on Nova's website under the e-book tab.

ENERGY POLICIES, POLITICS AND PRICES

U.S. EXPORTS OF OIL AND NATURAL GAS

ANALYSES OF SELECTED ISSUES

FRANK G. ADAMS
EDITOR

nova publishers
New York

For permission to use material from this book please contact us:
Telephone 631-231-7269; Fax 631-231-8175
Web Site: http://www.novapublishers.com

Library of Congress Cataloging-in-Publication Data

ISBN: 978-1-63463-904-0

Published by Nova Science Publishers, Inc. † New York

CONTENTS

PREFACE

This book examines what is known about price implications of removing crude oil export restrictions; other key potential implications; and implications of recent changes in market conditions on the SPR. This book also discusses and describes the status of applications to export liquefied natural gas-natural gas cooled to a liquid state for transport- and the Department of Energy's process to review them; and the status of applications to build LNG export facilities and FERC's process to review them.

Chapter 1 – Almost 4 decades ago, in response to the Arab oil embargo and recession it triggered, Congress passed legislation restricting crude oil exports and establishing the SPR to release oil to the market during supply disruptions and protect the U.S. economy from damage. After decades of generally falling U.S. crude oil production, technological advances have contributed to increasing U.S. production. Meanwhile, net crude oil imports—imports minus exports—have declined from a peak of about 60 percent of consumption in 2005 to 30 percent in the first 5 months of 2014. According to Energy Information Administration forecasts, net imports are expected to remain well below 2005 levels into the future. GAO was asked to provide information on the implications of removing crude oil export restrictions. This report examines what is known about (1) price implications of removing crude oil export restrictions; (2) other key potential implications; and (3) implications of recent changes in market conditions on the SPR. GAO reviewed four studies on crude oil exports, including two sponsored by industry, and summarized the literature and views of a nonprobability sample of stakeholders including academic, industry, and other experts.

Chapter 2 – Technological advances in hydraulic fracturing and horizontal drilling have resulted in a dramatic increase in the amount of natural gas that can be produced domestically. DOE is responsible for reviewing applications to export LNG—natural gas cooled to a liquid state for transport—and, under the Natural Gas Act, must approve an application unless it finds that approval is not consistent with the public interest. Since 2010, DOE has received 35 applications to export LNG that must address the public interest question. In addition, under NEPA, FERC is required to assess how LNG export facilities may affect the environment and is responsible for granting approval to build and operate export facilities. Since 2010, FERC has received 17 applications to construct export facilities. GAO was asked to report on the federal process for reviewing applications to export LNG. This report describes (1) the status of applications to export LNG and DOE's process to review them and (2) the status of applications to build LNG export facilities and FERC's process to review them. GAO reviewed laws, regulations, and guidance; examined export approvals; visited LNG facilities; and interviewed federal and state agency officials and industry representatives, including LNG export permit applicants. GAO is not making any recommendations in this report.

Chapter 3 – Recent growth in U.S. natural gas and crude oil production has fundamentally shifted the energy supply and demand balance in key U.S. energy markets. Many analysts and energy producers argue that there is currently an oversupply of natural gas in U.S. gas-producing regions, and that supplies of certain light sweet Western crudes (notably from the Bakken region) exceed the demand of Gulf Coast refineries. The result has been a spate of applications by the U.S. natural gas industry for federal permits to export natural gas to overseas buyers. Such exports are permitted by statute, subject to a national interest determination by the Department of Energy. Perceived oversupply of Bakken crude oil has similarly led to calls by some market stakeholders and policy makers for a relaxation of the federal restrictions on crude oil exports. Numerous legislative proposals in the 113th Congress, such as the Natural Gas Export Promotion Act of 2014 (S. 2494) and the Crude Oil Export Act (H.R. 4349), would facilitate increased exports of domestic energy resources. Whether—or to what extent—the United States should allow overseas exports of crude oil and natural gas is the subject of ongoing Congressional debate. However, U.S. exports of both natural gas and crude oil are not new. Congress and prior presidential administrations have authorized such exports before in circumstances where they were viewed as beneficial to the United States. The case of Alaska is of particular relevance because it is a major energy producer and has a history of both crude oil and natural gas exports going back several decades. Exports of crude oil from Alaska are currently authorized by three different statutes pertaining to (1) crude transported via the Trans Alaska Pipeline System, (2) Cook Inlet crude, and (3) crude oil exported to Canada under the North American Free Trade Agreement. Export of Cook Inlet natural gas from the Kenai LNG terminal has been authorized repeatedly since 1969. Export of natural gas from Alaska's North Slope was authorized in 1988, although the applicability of this authorization to the proposed Alaska LNG Project has not been resolved. In the context of the current export debate, a review of Alaskan crude oil and natural gas exports may offer historical perspectives of value to policy makers. Alaska's experience shows that— under a specific set of circumstances—such exports have been viewed as in the national interest both by Congress and successive presidential administrations. Regional production trends, market prices, and local commodity demand have all been important considerations in establishing export policies for Alaska. It is also instructive to note that, even with long-standing export approvals in effect, exports of crude oil and natural gas from Alaska have been relatively modest. Production economics and competitive market forces will be the ultimate determinant of export volumes where exports are permitted. In the context of broader U.S. oil and natural gas export policy the Alaska experience may raise several key questions. To what extent does the rationale for energy exports from Alaska—which is geographically isolated—apply to other U.S. supply regions? What is the interplay between overseas exports and the maritime shipping industry? What are the expectations for capital investment by developers to support export production and how might they affect the nation's overall oil and gas supplies? What are the environmental impacts of increased production for export? How might international trade agreements influence the U.S. government's ability to tailor its oil and natural gas export policies by region? As Congress continues its oversight of the nation's energy resources, understanding these issues may be important.

Chapter 4 – Recent technological developments have led to an increase in domestic production of natural gas and crude oil. As a result, there is interest among some parties in exporting liquefied natural gas (LNG) and crude oil to take advantage of international markets. This has placed new attention on the laws and regulations governing, and in many cases restricting, the export of fossil fuels. In most cases, export of fossil fuels requires federal authorization of both the act of exporting the fuel and the facility that will be employed to export the fuel. For example, the export of natural gas is permitted by the Department of Energy's Office of Fossil Energy, while the construction and operation of the export facility must be authorized by the Federal Energy Regulatory Commission (FERC). Oil exports are restricted, but an export that falls under one of several exemptions can be authorized by the Department of

Commerce's Bureau of Industry and Security. Oil pipelines that cross international borders must be permitted by the State Department. Coal exports do not require special authorization specific to the commodity; however, as with natural gas and crude oil, other generally applicable federal statutes and regulations may apply to the export of coal. Restrictions on exports of fossil fuels could potentially have implications under international trade rules. They may possibly be inconsistent with the most favored nation requirement of Article I of the General Agreement on Tariffs and Trade 1994 (GATT 1994) if certain World Trade Organization (WTO) members are treated differently than others. Limits on exports could also potentially violate the prohibition on export restrictions contained in Article XI of the GATT 1994 if they prescribe vague and unspecified criteria for export licensing. However, an export licensing regime does not appear to constitute a "subsidy" to downstream users of fossil fuels under WTO rules. Article XXI, the exception for essential security interests, may provide justification for potential violations of GATT Articles I and XI. The United States has traditionally considered this exception to be self-judging. However, it is possible that a panel or the Appellate Body might scrutinize the United States' use of the exception. Article XX of the GATT provides additional exceptions that a member country may invoke if it is found to be in violation of any GATT obligations. For example, WTO members may maintain an otherwise GATT inconsistent measure if it is necessary to protect an exhaustible natural resource or necessary to protect human health or the environment. Article XIII potentially requires that if an otherwise GATT inconsistent measure is permitted to remain in force due to an Article XX exception, the measure must be administered in a nondiscriminatory manner. Export restrictions that treat WTO members differently would appear not to satisfy the potential nondiscriminatory requirements of Article XIII.

In: U.S. Exports of Oil and Natural Gas
Editor: Frank G. Adams

ISBN: 978-1-63463-904-0

Chapter 1

CHANGING CRUDE OIL MARKETS: ALLOWING EXPORTS COULD REDUCE CONSUMER FUEL PRICES, AND THE SIZE OF THE STRATEGIC RESERVES SHOULD BE REEXAMINED*

United States Government Accountability Office

WHY GAO DID THIS STUDY

Almost 4 decades ago, in response to the Arab oil embargo and recession it triggered, Congress passed legislation restricting crude oil exports and establishing the SPR to release oil to the market during supply disruptions and protect the U.S. economy from damage. After decades of generally falling U.S. crude oil production, technological advances have contributed to increasing U.S. production. Meanwhile, net crude oil imports—imports minus exports—have declined from a peak of about 60 percent of consumption in 2005 to 30 percent in the first 5 months of 2014. According to Energy Information Administration forecasts, net imports are expected to remain well below 2005 levels into the future.

---

* This is an edited, reformatted and augmented version of a United States Government Accountability Office publication, No. GAO-14-807, dated September 2014.

GAO was asked to provide information on the implications of removing crude oil export restrictions. This report examines what is known about (1) price implications of removing crude oil export restrictions; (2) other key potential implications; and (3) implications of recent changes in market conditions on the SPR. GAO reviewed four studies on crude oil exports, including two sponsored by industry, and summarized the literature and views of a nonprobability sample of stakeholders including academic, industry, and other experts.

What GAO Recommends

In view of changing market conditions and in tandem with activities to assess other aspects of the SPR, GAO recommends that the Secretary of Energy reexamine the size of the SPR. In commenting on a draft of this report, DOE concurred with GAO's recommendation.

What GAO Found

The studies GAO reviewed and stakeholders interviewed suggest that removing crude oil export restrictions is likely to increase domestic crude oil prices but decrease consumer fuel prices. Prices for some U.S. crude oils are lower than international prices—for example, one benchmark U.S. crude oil averaged $101 per barrel in 2014, while a comparable international crude oil averaged $109. Studies estimate that U.S. crude oil prices would increase by about $2 to $8 per barrel—bringing them closer to international prices. At the same time, studies and some stakeholders suggest that U.S. prices for gasoline, diesel, and other consumer fuels follow international prices, so allowing crude oil exports would increase world supplies of crude oil, which is expected to reduce international prices and, subsequently, lower consumer fuel prices. Some stakeholders told GAO that there could be important regional differences in the price implications of removing crude oil export restrictions. Some stakeholders cautioned that estimates of the implications of removing export restrictions are uncertain due to several factors such as the extent of U.S. crude oil production increases, how readily U.S. refiners are able to absorb such increases, and how the global crude oil market responds to increasing U.S. production.

The studies GAO reviewed and stakeholders interviewed generally suggest that removing crude oil export restrictions may also have the following implications:

- **Crude oil production.** Removing export restrictions would increase domestic production—8 million barrels per day in April 2014—because of increasing domestic crude oil prices. Estimates range from an additional 130,000 to 3.3 million barrels per day on average from 2015 through 2035.
- **Environment.** Additional crude oil production may pose risks to the quality and quantity of surface groundwater sources; increase greenhouse gas and other emissions; and increase the risk of spills from crude oil transportation.
- **The economy.** Removing export restrictions is expected to increase the size of the economy, with implications for employment, investment, public revenue, and trade. For example, removing restrictions is expected to contribute to further declines in net crude oil imports, reducing the U.S. trade deficit.

Changing market conditions have implications for the size, location, and composition of Department of Energy's (DOE) Strategic Petroleum Reserve (SPR).

In particular, increased domestic crude oil production and falling net imports may affect the ideal size of the SPR. Removing export restrictions is expected to contribute to additional decreases in net imports in the future. As a member of the International Energy Agency, the United States is required to maintain public and private reserves of at least 90 days of net imports but, as of May 2014, the SPR held reserves of 106 days—worth about $73 billion—and private industry held reserves of 141 days.

DOE has taken some steps to assess the implications of changing market conditions on the location and composition of the SPR but has not recently reexamined its size. GAO has found that agencies should reexamine their programs if conditions change. Without such a reexamination, DOE cannot be assured that the SPR is sized appropriately and risks holding excess crude oil that could be sold to fund other national priorities.

ABBREVIATIONS

API	American Petroleum Institute
BIS	Bureau of Industry and Security
DOE	Department of Energy
EIA	Energy Information Administration
EPCA	Energy Policy and Conservation Act of 1975
IEA	International Energy Agency
OPEC	Organization of the Petroleum Exporting Countries
RFF	Resources for the Future
SPR	Strategic Petroleum Reserve
WTI	West Texas Intermediate

* * *

September 30, 2014

The Honorable Lisa Murkowski
Ranking Member
Committee on Energy and Natural Resources
United States Senate

Dear Senator Murkowski:

Almost 4 decades ago, Congress passed legislation restricting U.S. crude oil exports and establishing the Strategic Petroleum Reserve (SPR) in response to the Arab oil embargo and economic recession it triggered. In recent years, however, crude oil market conditions have changed, reversing decades-long trends in declining domestic crude oil production and increasing crude oil imports. Monthly crude oil production has increased by almost 68 percent from 2008 through April 2014, and increases in production in 2012 and 2013 were the largest annual increases since the beginning of U.S. commercial crude oil production in 1859, according to the Energy Information Administration (EIA).[1] With growing production, net crude oil imports—imports minus exports—have declined from a peak of about 60 percent of consumption in 2005 to about 30 percent in the first 5 months of 2014. In response, some members of Congress proposed legislation to remove crude oil export restrictions, while others argued that restrictions should remain in place and not be modified.[2] Recent congressional hearings stressed the need to understand how changing crude oil export restrictions could affect crude oil prices and the prices of consumer fuels refined from crude oil, such as gasoline and diesel.

At the same time, crude oil market changes may have implications for the SPR, the largest government-held emergency stockpile of crude oil in the world. The SPR holds 691 million barrels of crude oil in underground salt caverns along the Gulf Coast in Louisiana and Texas. In the event of a crude oil supply disruption, the SPR relies on the existing commercial distribution and refining system to transport and process crude oil into usable products for sale to the public. Increasing crude oil production, shrinking crude oil imports, changing crude oil and fuel distribution patterns, and an evolving U.S. refining industry could have important implications for the SPR.

You asked us to provide information on the implications of removing crude oil export restrictions. This report examines what is known about: (1) the potential effect of removing crude oil export restrictions on prices of crude oil and consumer fuels; (2) other potential implications of removing crude oil export restrictions; and (3) implications for the SPR, if any, from recent changes in crude oil market conditions.

To conduct this work, we reviewed information, including studies by federal agencies, consultants, and academics, and summarized the results of interviews from a nonprobability sample of 17 stakeholders. We identified relevant information by conducting a literature search and obtaining suggestions from stakeholders we interviewed.[3] We identified and summarized the results of four recent studies that estimate the potential implications of removing crude oil export restrictions. Two of these studies were sponsored by industry and conducted by consultants, another was sponsored by a research organization and conducted by consultants, and the fourth was conducted at a research organization.[4] To assess the reasonableness of these studies, we conducted a high-level review of the assumptions and methods used, interviewed the studies' authors, and obtained views of other stakeholders. We determined that these studies were reasonable for describing what is known about the range of potential implications but identified several limitations, which we discuss later in this report. We did not identify any other publicly available primary studies of the implications of removing crude oil export restrictions. Stakeholders included representatives of companies and interest groups with a stake in the outcome of decisions regarding crude oil export restrictions, as well as academic, industry, and other experts. We selected stakeholders based on our literature review and recommendations from agency officials and others. We asked the same questions during each interview but also discussed individual stakeholders' perspectives, as appropriate. We summarized their views, noting areas of consensus and disagreement. We may not have identified all stakeholders with a view on this topic, but we sought to balance the group with the range of perspectives. The views of stakeholders we selected are not generalizable to all potential stakeholders, but they provide illustrative examples of the range of views. To assess the implications of recent trends on the SPR, we interviewed Department of Energy (DOE) officials and reviewed agency and other documents. Data in this report are primarily from EIA, the International Energy Agency (IEA), and Bloomberg.[5] To assess the reliability of these data, we reviewed relevant documentation, interviewed EIA and Bloomberg officials, and compared the data with similar data published in other sources. We determined these data to be sufficiently reliable for the purposes of this report. Appendix I provides additional information on the four studies we reviewed, and appendix II lists the stakeholders we interviewed.

We conducted this performance audit from April 2014 to September 2014 in accordance with generally accepted government auditing standards. Those standards require that we plan and perform the audit to obtain sufficient, appropriate evidence to provide a reasonable basis for our findings and conclusions based on our audit objectives. We believe that the evidence obtained provides a reasonable basis for our findings and conclusions based on our audit objectives.

BACKGROUND

This section describes crude oil export restrictions, the SPR, and recent trends in U.S. crude oil production and the petroleum refining industry.

Crude Oil Export Restrictions

The export of domestically produced crude oil has generally been restricted since the 1970s. In particular, the Energy Policy and Conservation Act of 1975 (EPCA) led the Department of Commerce's Bureau of Industry and Security (BIS) to promulgate regulations which require crude oil exporters to obtain a license.[6] These regulations provide that BIS will issue licenses for the following crude oil exports:

- exports from Alaska's Cook Inlet,
- exports to Canada for consumption or use therein,
- exports in connection with refining or exchange of SPR crude oil,
- exports of certain California crude oil up to 25,000 barrels per day,
- exports consistent with certain international energy supply agreements,
- exports consistent with findings made by the President under certain statutes, and
- exports of foreign origin crude oil that has not been commingled with crude oil of U.S. origin.

Other than for these exceptions, BIS considers export license applications for exchanges involving crude oil on a case-by-case basis, and BIS can approve them if it determines that the proposed export is consistent with the national interest and purposes of EPCA.[7] In addition to BIS's export controls, other statutes control the export of domestically produced crude oil depending on where it was produced and how it is transported.[8] In these cases, BIS can approve exports only if the President makes the necessary findings under applicable laws.[9] Some of the authorized exceptions, outlined above, are the result of such Presidential findings.

According to NERA, no other major oil producing country currently restricts crude oil exports.[10]

BIS approved about 30 to 40 licenses to export domestic crude oil per year from fiscal years 2008 through 2010. The number of BIS approved licenses increased to 103 in fiscal year 2013. Meanwhile, crude oil exports increased from less than 30 thousand barrels per day in 2008 to 396 thousand barrels per day in June 2014—the highest level of exports since 1957. Nearly all domestic crude oil exports have gone to Canada.

The SPR

To help protect the U.S. economy from damage caused by crude oil supply disruptions, Congress authorized the SPR in 1975. The SPR is owned by the federal government and operated by DOE. The SPR is authorized to hold up to 1 billion barrels of crude oil and has the capacity to store 727 million barrels of crude oil in salt caverns located at sites in Texas and Louisiana. According to DOE, the SPR held crude oil valuded at almost $73 billion dollars as of May, 2014. From fiscal year 2000 through 2013, the federal government spent about $0.5 billion to purchase crude oil, and spent $2.5 billion for operations and maintenance of the reserve.

The United States is a member of the IEA and has agreed, along with 28 other member nations, to maintain reserves of crude oil or petroleum products equaling 90 days of net imports and to release these reserves and reduce demand during oil supply disruptions.[11] The 90-day reserve requirement can be made up of government reserves, such as the SPR, and inventory reserves held by private industry.[12]

Under conditions prescribed by the Energy Policy and Conservation Act, as amended, the President and the Secretary of Energy have discretion to authorize the release of crude oil from the SPR to minimize significant supply disruptions.[13] In the event of a crude oil supply disruption, the SPR can supply the market by selling stored crude oil or trading crude oil in exchange for an equal quantity of crude oil plus an additional amount as a premium to be returned to the SPR in the future. When crude oil is released from the SPR, it flows through commercial pipelines or on waterborne vessels to refineries, where it is converted into gasoline and other petroleum products, and then transported to distribution centers for sale to the public.

Trends in U.S. Crude Oil Production and the Petroleum Refining Industry

Reversing a decades-long decline, U.S. crude oil production has increased in recent years. According to EIA data, U.S. production of crude oil reached its highest level in 1970 and generally declined through 2008, reaching a level of almost one-half of its peak. During this time, the United States increasingly relied on imported crude oil to meet growing domestic energy needs. However, recent improvements in technologies have allowed producers to extract crude oil from shale formations that were previously considered to be inaccessible because traditional techniques did not yield sufficient amounts for economically viable production. In particular, the application of horizontal drilling techniques and hydraulic fracturing—a process that injects a combination of water, sand, and chemical additives under high pressure to create and maintain fractures in underground rock formations that allow crude oil and natural gas to flow—have increased U.S. crude oil and natural gas production.[14] Monthly domestic crude oil production has increased from an average of about 5 million barrels per day in 2008 to about 8.4 million barrels per day in April 2014, an increase of almost 68 percent.

As we previously found, the growth in U.S. crude oil production has lowered the cost of some domestic crude oils.[15] For example, prices for West Texas Intermediate (WTI) crude oil—a domestic crude oil used as a benchmark for pricing—was historically about the same price as Brent, an international benchmark crude oil from the North Sea between Great Britain and the European continent.[16] However, from 2011 through June 13, 2014, the price of WTI averaged $14 per barrel lower than Brent (see fig. 1). In 2014, prices for these benchmark crude oils narrowed somewhat, and WTI averaged $101 through June 13, 2014, while Brent averaged $109. The development of U.S. crude oil production has created some challenges for crude oil transportation infrastructure because some production has been in areas with limited linkages to refining centers. According to EIA, these infrastructure constraints have contributed to discounted prices for some domestic crude oils.

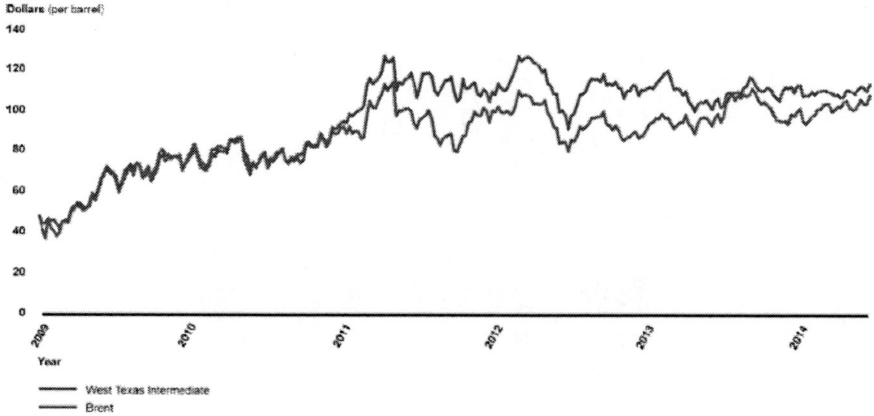

Source: GAO analysis of Bloomberg data. | GAO-14-807.

Note: West Texas Intermediate is a domestic crude oil used as a benchmark for pricing, and Brent is an international benchmark from the North Sea between Great Britain and the European continent.

Figure 1. Weekly West Texas Intermediate and Brent Crude Oil Prices, 2009-June 2014.

Much of the crude oil currently produced in the United States has characteristics that differ from historic domestic production. Crude oil is generally classified according to two parameters: density and sulfur content. Less dense crude oils are known as "light," while denser crude oils are known as "heavy." Crude oils with relatively low sulfur content are known as "sweet," while crude oils with higher sulfur content are known as "sour." As shown in figure 1, according to EIA, production of new domestic crude oil has tended to be light oils. Specifically, according to EIA estimates about all of the 1.8 million barrels per day growth in production between 2011 and 2013 consisted of lighter sweet crude oils.[17] EIA also forecasts that lighter crude oils will make up a significant portion of production growth in 2014 and 2015—about 60 percent.

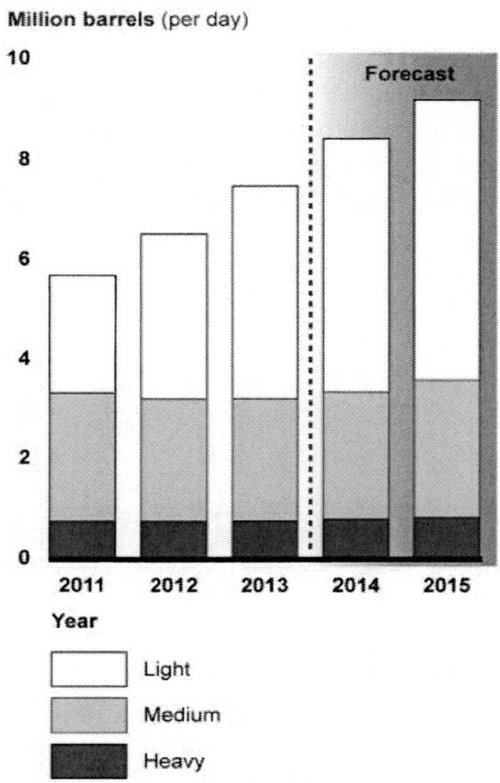

Source: GAO analysis of Energy Information Administration data. I GAO-14-807.

Note: The density, or gravity, of a crude oil is specified using the American Petroleum Institute (API) gravity standard, which measures the weight of crude oil in relation to water, which has an API gravity of 10 degrees. Heavy crude oils include those with an API gravity of less than 27; medium includes crude oil with an API from 27 to 35; and light includes crude oil with API gravities of 35 and above.

Figure 2. U.S. Crude Oil Production and Energy Information Administration Forecast of Production by Crude Oil Type, 2011-2015.

Light crude oil differs from the crude oil that many U.S. refineries are designed to process. Refineries are configured to produce transportation fuels and other products (e.g., gasoline, diesel, jet fuel, and kerosene) from specific types of crude oil. Refineries use a distillation process that separates crude oil into different fractions, or interim products, based on their boiling points, which can then be further processed into final products. Many refineries in the United States are configured to refine heavier crude oils, and have therefore been able to take advantage of historically lower prices of heavier crude oils.[18] For example, in 2013, the average density of crude oil used at domestic refineries was 30.8 while nearly all of the increase in production in recent years has been lighter crude oil with a density of 35 or above.

According to EIA, additional production of light crude oil over the past several years has been absorbed into the market through several mechanisms, but the capacity of these mechanisms to absorb further increases in light crude oil production may be limited in the future as follows:

- **Reduced imports of similar grade crude oils**: According to EIA, additional production of light oil in the past several years has primarily been absorbed by reducing imports of similar grade crude oils. Light crude oil imports fell from 1.7 million barrels per day in 2011 to 1 million barrels per day in 2013. There may be dwindling amounts of light crude oil imports that can be reduced in the future, according to EIA.
- **Increased crude oil exports**: As discussed above, crude oil exports have increased recently, from less than 30 thousand barrels per day in 2008 to 396 thousand barrels per day in June 2014. Continued increases in crude oil exports will depend, in part, on the extent of any relaxation of current export restrictions, according to EIA.
- **Increased use of light crude oils at domestic refineries**: Domestic refineries have increased the average gravity of crude oils that they refine. The average API gravity of crude oil used in U.S. refineries increased from 30.2 degrees in 2008 to 30.8 degrees in 2013. Continued shifts to use additional lighter crude oils at domestic refineries can be enabled by investments to relieve constraints associated with refining lighter crude oils at refineries that were optimized to refine heavier crude oils.
- **Increased use of domestic refineries**: In recent years, domestic refineries have been run more intensively, allowing the use of more domestic crude oils. Utilization—a measure of how intensively

refineries are used that is calculated by dividing total crude oil and other inputs used at refineries by the amount refineries can process under usual operating conditions—increased from 86 percent in 2011 to 88 percent in 2013. There may be limits to further increases in utilization of refineries that are already running at high rates.

REMOVING CRUDE OIL EXPORT RESTRICTIONS IS EXPECTED TO INCREASE DOMESTIC CRUDE OIL PRICES AND COULD DECREASE CONSUMER FUEL PRICES

The studies we reviewed and stakeholders we interviewed generally suggest some domestic crude oil prices would increase if crude oil export restrictions were removed, while consumer fuel prices could decrease, although the extent of consumer fuel price changes are uncertain and may vary by region.

Some Domestic Crude Oil Prices Are Expected to Increase

Studies we reviewed and most of the stakeholders we interviewed suggest that some domestic crude oil prices would increase if crude oil export restrictions were removed.[19] As discussed above, increasing domestic crude oil production has resulted in lower prices of some domestic crude oils compared with international benchmark crude oils.[20] Three of the studies we reviewed also said that, absent changes in crude oil export restrictions, the expected growth in crude oil production may not be fully absorbed by domestic refineries or through exports (where allowed), contributing to even wider differences in prices between some domestic and international crude oils.[21] By removing the export restrictions, these domestic crude oils could be sold at prices closer to international prices, reducing the price differential and aligning the price of domestic crude oil with international benchmarks. While the studies we reviewed and most of the stakeholders we interviewed agree that domestic crude oil prices would increase if crude oil export restrictions were removed, stakeholders highlighted several factors that could affect the extent of price increases. The studies we reviewed made assumptions about these factors, and actual price implications of removing crude oil export restrictions may differ from those estimated in the studies depending on how export restrictions and market conditions evolve. Specifically, stakeholders raised the following three key uncertainties:

- **Extent of future increases in crude oil production**. As we recently found, forecasts anticipate increases in domestic crude oil production in the future, but the projections are uncertain and vary widely.[22] Two of the studies and two stakeholders told us that, in the absence of exports, higher production of domestic light sweet crude oil would tend to increase the mismatch between such crude oils and the refining industry. In turn, one study indicated that a greater increase in production would increase the price effects of removing crude oil export restrictions. On the other hand, lower than anticipated production of such crude oil would lower potential price effects as the additional crude oil could more easily be absorbed domestically.
- **Extent to which crude oil production increases can be absorbed**. The domestic refining industry and exports to Canada have absorbed the increases in domestic crude oil production thus far, and one stakeholder told us the domestic refining industry could provide sufficient capacity to absorb additional future crude oil production. This stakeholder said that refineries have the capacity to refine another 400,000 barrels a day of light crude oil, some of which is not being used because of infrastructure or logistics constraints. The industry is planning to develop or is in the process of developing the capacity to process an additional 500,000 barrels a day of light crude oil, according to this stakeholder. The current capacity that is not being utilized plus capacity that is planned or in development would constitute a total capacity to refine 900,000 barrels per day of light crude oil. To the extent that light crude oil production increases by less than this amount, the gap in prices between WTI and Brent could close in the future as increased crude oil supplies are absorbed.[23] This would reduce the extent to which domestic crude oil prices increase if crude oil export restrictions are removed. On the other hand, some stakeholders suggested that the U.S. refining industry will not be able to keep pace with increasing U.S. light crude oil production. For example, IHS stated that refinery investments to process additional light crude oil face significant risks in the form of potentially stranded investments if export restrictions were to change, and this could result in investments not being made as quickly as anticipated.

- **Extent to which export restrictions change.** Aspects of the export restrictions could be further defined or interpreted in ways that could change the pricing dynamics of domestic crude oil markets. Recently, two companies received clarification from the Department of Commerce that condensate—a type of light crude oil[24]—that has been processed through a distillation tower is not considered crude oil and so not subject to export restrictions.[25] One stakeholder stated that this may lead to more condensate exports than expected.[26]

Within the context of these uncertainties, estimates of potential price effects vary in the four studies we reviewed, as shown in table 1. Specifically, estimates in these studies of the increase in domestic crude oil prices due to removing crude oil export restrictions range from about $2 to $8 per barrel.[27] For comparison, at the beginning of June 2014, WTI was $103 per barrel, and these estimates represent 2 to 8 percent of that price. In addition, NERA found that removing export restrictions would have no measurable effect in a case that assumes a low future international oil price of $70 per barrel in 2015 rising to less than $75 by 2035. According to NERA, current production costs are close to these values, so that removing export restrictions would provide little incentive to produce more light crude oil.

Table 1. Crude Oil Price Implications of Removing Crude Oil Export Restrictions from Four Studies

	Resources for the Future	ICF International	IHS	NERA
U.S. crude oil price	Midwest refiner acquisition costs increase $6.68 per barrela	WTI prices $2.35 to $4.19 per barrel higher on average from 2015-2035	$7.89 per barrel higher on average from 2016-2030	Prices increase $1.74 per barrel in the reference case and $5.95 per barrel in the high case on average from 2015-2035. b

Sources: GAO analysis of Resources for the Future, ICF International, IHS, and NERA studies. | GAO-14-807

Note: Estimates are in 2014 year dollars.

[a]Refiner acquisition costs are the costs of crude oil including transportation and other fees paid by the refiner. Such costs may be closely related to the prices of crude oil discussed in this report.

[b]Implications refer to the difference between the reference case and its baseline with export restrictions in place, and the difference between the high oil and gas recovery case and its corresponding baseline. NERA also found that removing crude oil export restrictions would have no measurable effect in the low world oil price case.

Consumer Fuel Prices Could Decrease, but Effects May Vary by Region

The studies we reviewed and most of the stakeholders we interviewed suggest that consumer fuel prices, such as gasoline, diesel, and jet fuel, could decrease as a result of removing crude oil export restrictions. A decrease in consumer fuel prices could occur because they tend to follow international crude oil prices rather than domestic crude oil prices, according to the studies and most of the stakeholders. If domestic crude oil exports caused international crude oil prices to decrease, consumer fuel prices could decrease as well.[28] Table 2 shows that the estimates of the price effects on consumer fuels vary in the four studies we reviewed. Price estimates range from a decrease of 1.5 to 13 cents per gallon. These estimates represent 0.4 to 3.4 percent of the average U.S. retail gasoline price at the beginning of June 2014. In addition, NERA found that removing export restrictions has no measurable effect on consumer fuel prices in a case that assumes a low future world crude oil price.

Table 2. Consumer Fuel Price Implications of Removing Crude Oil Export Restrictions from Four Studies

	Resources for the Future	ICF International	IHS	NERA[a]
U.S. Consumer Fuel Prices	Gasoline prices decline by 1.8 to 4.6 center per gallon on average	Petroleum product prices decline by 1.5 to 2.4 cents per gallon on average from 2015-2035	Gasoline prices decline by 9 to 13 cents per gallon on average from 2016-2030	Petroleum product prices decline by 3 cents per gallon on average from 2015-2035 in the reference case and 11 cents per gallon in the high case. Gasoline prices decline by 3 cents per gallon in the reference case and 10 cents per gallon in the high case.

Sources: GAO analysis of Resources for the Future, ICF International, IHS, and NERA studies. | GAO-14-807

Note: Dollar estimates are in 2014 year dollars.

[a]Implications refer to the difference between the reference case and its baseline with export restrictions in place, and the difference between the high oil and gas recovery case and its corresponding baseline. NERA also found that removing crude oil export restrictions has no measurable effect in the world crude oil low price case.

Price Effects of Allowing Alaskan North Slope Crude Oil Exports

In 1995, Congress removed the restrictions on the export of Alaskan North Slope crude oil. From the time the restrictions were removed until 2004, about 2.7 percent of Alaskan North Slope crude oil was exported; however, no Alaskan North Slope crude oil has been exported since 2004.The experience of allowing Alaskan North Slope crude oil exports may illustrate some of the potential effects of removing crude oil export restrictions nationally. In 1999, we reviewed the effects of allowing Alaskan North Slope crude oil exports and concluded that:[a]

- lifting the export ban raised the relative prices of Alaskan North Slope and comparable California crude oils by between $0.98 and $1.30 per barrel;[b]
- some refiners' costs increased commensurate with the increase in crude oil prices; and
- consumer fuel prices for gasoline, diesel, and jet fuel did not increase.

The effect of removing the export restrictions for Alaskan North Slope oil is not completely understood due to data limitations and the difficulty of separating the effects of removing the export restrictions from other market changes that occurred at the same time.

Source: GAO. | GAO-14-807

[a]GAO, Alaskan North Slope Oil: Limited Effects of Lifting Export Ban on Oil and Shipping Industries and Consumers, GAO/RCED-99-191 (Washington, D.C., July 1, 1999).

[b]These estimates have not been adjusted for inflation.

The effect of removing crude oil export restrictions on domestic consumer fuel prices depends on several uncertain factors. First, it depends on the extent to which domestic versus international crude oil prices determine the domestic price of consumer fuels. Recent research examining the relationship between domestic crude oil and gasoline prices concluded that low domestic crude oil prices in the Midwest during 2011 did not result in lower gasoline prices in that region.[29] This research supports the assumption made in all of the studies we reviewed that to some extent higher prices of some domestic crude oils as a result of removing crude oil export restrictions would not be passed on to consumer fuel prices. However, some stakeholders told us that this may not always be the case and that more recent or detailed data could show that lower prices for some domestic crude oils have influenced consumer fuel prices.

Second, the extent to which domestic consumer fuel prices could decline also depends on how the global crude oil market responds to the domestic crude oil entering the market. In this regard, stakeholders highlighted several uncertainties. In particular, the response of the Organization of the Petroleum Exporting Countries (OPEC) could have a large influence on any international crude oil price changes. The projections in the RFF, IHS, and ICF International studies assumed that OPEC would not respond by attempting to counterbalance the effect of increased U.S. exports by reducing its countries' exports. However, OPEC could seek to maintain international crude oil prices by pulling crude oil from the global market. In this case, the international crude oil price would not be affected by removing export restrictions, and consumer fuel prices would not decline. On the other hand, OPEC could increase production to maintain its large market share, which would push international crude oil prices and consumer fuel prices downward. NERA examined two alternative OPEC response cases, and found that gasoline prices would not generally be affected if OPEC reduces production, and that consumer fuel prices would decrease further if OPEC maintains its production in the face of lower global crude oil prices. In addition, one stakeholder questioned whether international crude oil prices would be affected by U.S. crude oil exports. Given the size of the global crude oil market, this stakeholder suggested that U.S. exports would have little to no effect on international crude oil prices.

Third, two of the stakeholders we interviewed suggested that there could be important regional differences in consumer fuel price implications, and that prices could increase in some regions—particularly the Midwest and the Northeast—due to changing transportation costs and potential refinery closures. For example, two stakeholders told us that because of requirements to use more expensive U.S.-built, -owned, and -operated ships to move crude oil between U.S. ports, allowing exports could enable some domestic crude oil producers to ship U.S. crude oil for less cost to refineries in foreign countries.[30] Specifically, representatives of one refiner told us that, if exports restrictions were removed, they could ship oil to their refineries in Europe at a lower cost than delivering the same oil to a refinery on the U.S. East Coast. According to another stakeholder, this could negatively affect the ability of some domestic refineries to compete with foreign refineries. Additionally, because refineries are currently benefiting from low domestic crude oil prices, some studies and stakeholders noted that refinery margins could be reduced if removing export restrictions increased domestic crude oil prices. As a result, some refineries could face an increased risk of closure, especially those located in the Northeast. As EIA reported in 2012, refinery closures in the Northeast could be associated with higher consumer fuel prices and possibly higher price volatility.[31] However, according to one stakeholder, domestic refiners still have a significant cost advantage in the form of less expensive natural gas, which is an important energy source for many refineries. For this and other reasons, one stakeholder told us they did not anticipate refinery closures as a result of removing export restrictions.

Removing Crude Oil Export Restrictions Is Expected to Increase Domestic Production and Have Other Effects

The studies we reviewed and stakeholders we interviewed generally suggest that removing crude oil export restrictions would increase domestic crude oil production and may affect the environment and the economy.

Removing Export Restrictions Is Expected to Increase Domestic Production, but Projections Vary

Studies we reviewed and stakeholders we interviewed generally agree that removing crude oil export restrictions would increase domestic crude oil production. Monthly domestic crude oil production has increased by almost 68 percent since 2008—from an average of about 5 million barrels per day in 2008 to 8.3 million barrels per day in April 2014. Even with current crude oil export restrictions, given various scenarios, EIA projects that domestic production will continue to increase and could reach 9.6 million barrels per day by 2019.[32] If export restrictions were removed, according to the four studies we reviewed, the increased prices of domestic crude oil are projected to lead to further increases in crude oil production. Projections of this increase varied in the studies we reviewed—from a low of an additional 130,000 barrels per day on average between 2015 and 2035, according to the ICF International study, to a high of an additional 3.3 million barrels per day on average between 2015 and 2035 in NERA's study.[33] This is equivalent to 1.5 percent to almost 40 percent of production in April 2014.

One stakeholder we spoke with told us that, although domestic demand for crude oil is not expected to change, production will rise as a result of increased international demand, primarily from Asia. For example, according to EIA, India was the fourth-largest consumer of crude oil and petroleum products in the world in 2013, and the country's dependence on imported crude oil continues to grow.[34] Another stakeholder stated that removing export restrictions could lead to increased local and regional opposition to crude oil production if the crude oil was primarily for export, which could affect domestic production.

Removing Export Restrictions Is Expected to Affect the Environment

Two of the studies we reviewed and most stakeholders we spoke with stated that the increased crude oil production that would result from removing the restrictions on crude oil exports may affect the environment. In September 2012, we found that crude oil development may pose certain inherent environmental and public health risks; however, the extent of the risk is unknown, in part, because the severity of adverse effects depend on various location- and process-specific factors, including the location of future shale oil and gas development and the rate at which it occurs, as well as geology, climate, business practices, and regulatory and enforcement activities.[35] The stakeholders who raised concerns identified the following risks related to crude oil production, about which GAO has reported in the past:

- **Water quality and quantity:** Increased crude oil production, particularly from shale, could affect the quality and quantity of surface and groundwater sources, but the magnitude of such effects is unknown. In October 2010, we found that water is needed for a number of oil shale development activities, including constructing facilities, drilling wells, generating electricity for operations, and reclamation of drill sites.[36] In 2012, we found that shale oil and gas development may pose a risk to surface water and groundwater because withdrawing water from streams, lakes, and aquifers for drilling and hydraulic fracturing could adversely affect water resources.[37] For example, we found that groundwater withdrawal could affect the amount of water available for other uses, including public and private water supplies. One of the stakeholders we interviewed suggested that water withdrawal is already an important consideration, particularly for areas experiencing drought. For example, the stakeholder noted that crude oil production and associated water usage already has implications for the Edwards Aquifer, a groundwater system serving the agricultural, industrial, recreational, and domestic needs of almost two million users in south central Texas. In addition, removing export restrictions may affect water quality. Another stakeholder told us that allowing crude oil exports would lead to more water pollution as a result of increased production through horizontal drilling.

- **Air quality:** Increased crude oil production may increase greenhouse gases and other air emissions because the use of consumer fuels would increase, and also because the crude oil production process often involves the direct release of pollutants into the atmosphere (venting) or burning fuels (flaring).[38] Two stakeholders told us that venting and flaring has escalated in North Dakota, in part because regulatory oversight and infrastructure have not kept pace with the recent surge in crude oil production in the state. In January 2014, the North Dakota Industrial Commission reported that nearly 30 percent of all natural gas produced in the state is flared. According to a 2013 report from Ceres, flaring in North Dakota in 2012 resulted in greenhouse gas emissions equivalent to adding 1 million cars to the road.[39] Another stakeholder told us that allowing crude oil exports would lead to more air pollution as a result of increased production through horizontal drilling and hydraulic fracturing. RFF estimated the potential environmental effect of removing export restrictions, estimating that increases in crude oil production and consumption would increase carbon dioxide emissions worldwide by almost 22 million metric tons per year.[40] By comparison, U.S. emissions from energy consumption totaled 5,393 million metric tons in 2013 according to EIA. NERA estimated that increased crude oil production and use of fossil fuels would increase greenhouse gas emissions by about 12 million metric tons of carbon dioxide equivalents per year on average from 2015 through 2035.[41]
- **Transportation challenges:** Increased crude oil production could exacerbate transportation challenges. In March 2014, we found that domestic and Canadian crude oil production has created some

challenges for U.S. crude oil transportation infrastructure. Some of the growth in crude oil production has been in areas with limited transportation to refining centers.[42] To address this challenge, refiners have relied on rail to transport crude oil. According to data from the Surface Transportation Board, rail moved about 236,000 carloads of crude oil in 2012, which is 24 times more than the roughly 9,700 carloads moved in 2008. As we recently found, as the movement of crude oil by rail has increased incidents such as spills and fires involving crude oil trains have also increased—from 8 incidents in 2008 to 119 incidents in 2013 according to Department of Transportation data.[43] Some stakeholders told us that removing export restrictions would increase the risk for crude oil spills by rail and other modes of transportation such as tankers. On the other hand, one stakeholder suggested that removing export restrictions could reduce the amount of crude oil transported by rail, in some instances, since the most economic way to export crude oil is by pipeline to a tanker. As a result, the number of rail accidents involving crude oil spills could decrease.

Removing Export Restrictions Is Expected to Affect the Economy

The studies we reviewed suggest that removing crude oil export restrictions would increase the size of the economy. Three of the studies project that removing export restrictions would lead to additional investment in crude oil production and increases in employment. This growth in the oil sector would—in turn—have additional positive effects in the rest of the economy.[44] For example, NERA projects an average of 230,000 to 380,000 workers would be removed from unemployment through 2020 if export restrictions were eliminated in 2015.[45] These employment benefits largely disappear if export restrictions are not removed until 2020 because by then the economy will have returned to full employment. Potential implications for investment, public revenue, and trade are as follows:

- **Investments:** According to one of the studies we reviewed, removing export restrictions may lead to more investment in crude oil exploration and production, but this investment could be somewhat offset by less investment in the refining industry. As discussed previously, removing export restrictions is expected to increase domestic crude oil production. Private investment in drilling rigs, engineering services, and transportation and logistics facilities, for example, is needed to increase domestic crude oil production. According to IHS, this will directly benefit industries such as machinery, fabricated metals, steel, chemicals, and engineering services. At the same time, removing export restrictions may decrease investment in the refining industry because the industry would not need extensive additional investment to accommodate lighter crude oils. For example, one stakeholder told us that, under current export restrictions, refining additional light crude oils may require capital investment to remove processing constraints at refineries that are designed to process heavier crude oils. Officials from one refining company told us that they had invested a significant amount of capital to refine lighter oils. For example, the refinery installed two new distillation towers to process lighter crude oils at a cost of $800 million. Such investments may not be necessary if export restrictions were removed.

- **Public revenue:** Two of the studies we reviewed suggest that removing export restrictions would increase government revenues, although the estimates of the increase vary. One study estimated that total government revenue would increase by a combined $1.4 trillion in additional revenue from 2016 through 2030 while another study estimated that U.S. federal, state, and local tax receipts combined with royalties from drilling on federal lands could increase by an annual average of $3.9 to $5.7 billion from 2015 through 2035.
- **Trade:** According to the studies we reviewed, removing export restrictions would contribute to further declines in net petroleum (i.e., crude oil, consumer fuels, and other petroleum products) imports and reduce the U.S. trade deficit. Three of the studies we reviewed estimated the effect of removing export restrictions on net petroleum imports, with ICF projecting a decline in net imports of about 100,000 to 300,000 barrels per day; IHS projecting a decline, but not providing a specific estimate; and NERA projecting a decline of about 0.6 to 3.2 million barrels per day. Further, according to one study, removing export restrictions could also improve the U.S. trade balance because the light sweet crude oils are usually priced higher than heavy, sour crude oils. One study estimated that removing export restrictions could improve the trade balance (narrow the U.S. trade deficit) by $8

to $15 billion per year on average from 2015 through 2035. Another study estimated that removing crude oil export restrictions would improve the trade balance by $72 to $101 billion per year from 2016 through 2030.

CHANGING MARKET CONDITIONS RAISE QUESTIONS ABOUT THE SIZE, LOCATION, AND COMPOSITION OF THE SPR

Changing market conditions—most importantly the significant increase in domestic production of crude oil from shale—have implications for the role of the SPR, including its appropriate size, location, and composition. DOE has taken some steps to reexamine the location and composition of the SPR in light of these changes, but has not recently reexamined its size.

Changing Crude Oil Market Conditions Affect the SPR

Recent and expected changes in crude oil markets have important implications for the role of the SPR, including its size, location, and composition.

DOE has recognized that recent increases in domestic crude oil production and correlating reductions in crude oil imports have changed how crude oil is transported around the United States, and that these changes carry potential implications for the operation and maintenance of the SPR. As discussed above, removing crude oil export restrictions would be expected to increase domestic crude oil production and contribute to further declines in net imports. Our review of DOE documents, prior GAO work, and discussions with stakeholders highlight three primary implications for the SPR.

Size

Increased domestic crude oil production and falling net petroleum imports may affect the ideal size of the SPR—how much the SPR should hold to optimize the benefits of protecting the economy from damage with the costs of holding the reserves.

One measure of the economy's vulnerability to oil supply disruptions is to assess net petroleum imports— imports minus exports petroleum imports have declined from a peak of 60 percent of consumption in 2005 to about 30 percent in the first half of 2014.

In 2006, net imports were expected to increase in the future, increasing the country's reliance on foreign crude oil. However, imports have declined and, according to EIA's most recent forecast, are expected to remain well below 2005 import levels into the future. (See fig. 3.) As discussed above, removing crude oil export restrictions would be expected to contribute to additional decreases in net petroleum imports in the future.

To the extent that changes in net imports reflect changes in vulnerability, these and other changes in the economy may have reduced the nation's vulnerability to supply disruptions. For example, a recent report by the President's Council of Economic Advisers suggests that decreased domestic petroleum demand, increased domestic crude oil production, more fuel efficient vehicles, and increased use of biofuels, have each contributed to reducing the vulnerability of the nation's economy to international crude oil supply disruptions.[46]

Although international crude oil supply and price volatility remains a risk, the report suggests that additional reductions in net petroleum imports could reduce those risks in the future. In addition, the SPR currently holds oil in excess of international obligations. As a member of the IEA, the United States is required to maintain reserves of crude oil or petroleum products equaling at least 90 days of net imports, which it does with a combination of public and private reserves.

According to IEA, as of May 2014, the SPR held 106 days of net imports, and private reserves held an additional 141 days of imports for a total of 247 days—well above the 90 days required by the IEA.[47] In light of these factors, some of the stakeholders we interviewed raised questions about whether such a large SPR is needed in the future. For example, one stakeholder indicated that SPR crude oil is surplus and no longer needed to protect the economy. However, other stakeholders highlighted the importance of maintaining the SPR. For example, one stakeholder said that the SPR should be maintained at the current level, and another said that the SPR serves an important "energy insurance" service.

DOE officials and one other stakeholder highlighted that, in addition to net imports, there are other factors that may affect the appropriate size of the SPR.[48]

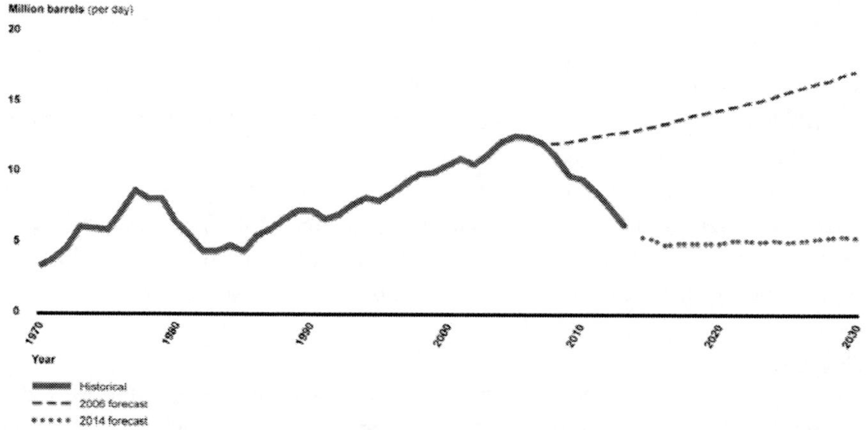

Source: GAO analysis of Energy Information Administration data and forecasts. | GAO-14-807.

Note: Forecast data are the 2006 and 2014 Annual Energy Outlook reference scenarios.

Figure 3. U.S. Net Petroleum Imports, Historical (1970-2013) and Energy Information Administration Forecasts.

Location

According to DOE, changes in how crude oil is transported throughout the United States and in the existing infrastructure surrounding SPR facilities have implications for the location of the SPR. Crude oil in the SPR is stored along the Gulf Coast, where it can take advantage of being in close proximity to a major refining center, as well as distribution points for tankers, barges, and pipelines that can carry crude oil from the SPR to refineries in other regions of the country.

Most of the system of crude oil pipelines in the United States was constructed in the 1950s, 1960s, and 1970s to accommodate the needs of the refining sector and demand centers at the time. According to DOE officials, the existing infrastructure was designed primarily to move crude oil from the southern United States to the North. The SPR has historically been able to rely on this distribution system to reach a large portion of the nation's refining capacity. But, with increases in crude oil production in the Northern U.S. and imports of crude oil from Canada, the distribution system has changed to increase crude oil flows south to the Gulf Coast. Such changes include new pipeline construction and expansions, flow reversals in existing pipelines, and increased utilization of terminals and marine facilities.

Such changes may make it more difficult to move crude oil from the SPR to refineries in certain regions of the United States, such as the Midwest, where almost 20 percent of the nation's refining capacity is located, according to EIA data. Some stakeholders raised questions about the location of the SPR. One stakeholder also suggested that holding SPR crude oil in the western United States may better ensure access to crude oil in the case of a disruption, since the West has no pipeline connectivity to the Gulf Coast. According to DOE, recent changes to crude oil distribution in the United States could have significant implications for the operation and maintenance of the SPR.

Composition

In 2006, we reported that the type of crude oil in the SPR was not compatible with all U.S. refineries. We reported that some U.S. refineries processed crude oils heavier than those stored in the SPR. We found that in the event of a disruption in the supply of heavy crude oil, refineries configured to use heavy crude oil would not be able to efficiently refine crude oil from the SPR and would likely reduce production of some petroleum products. As we reported, in such instances, prices for heavy crude oil products could increase, reducing the SPR's effectiveness to limit economic damage.[49] Refinery officials we spoke with noted that the SPR should contain heavier crude oils that domestic refineries could refine in the event of a supply disruption. Since our 2006 report, domestic production of light sweet crude oil has increased. According to EIA, roughly 96 percent of the 1.8 million-barrel per day growth in production from 2011 to 2013 consisted of light sweet grades with API gravity of 40 or above. As a result, imports of light crude oils have declined, and U.S. reliance on imported heavy crude oil has increased from 37 percent of total imports in 2008 to 50 percent of total imports in 2013, as shown in figure 4.

However, DOE officials raised concerns about the prospect of storing additional heavy crude oil in the SPR. According to DOE officials and a 2010 report by DOE, storing heavy crude oil in the SPR would limit the SPR's ability to respond to nonheavy crude oil disruptions, such as a loss of Middle East medium sour crude oils. In addition, storing more heavy crude would require infrastructure improvements.[50] At the same time, DOE officials also stated that, based on recent conversations with refinery officials, no U.S. refineries would have difficulty using SPR crude oils. Another issue raised by some stakeholders we interviewed is that the SPR holds primarily crude oil, and some stakeholders told us that holding additional consumer fuels could be beneficial.[51]

Many recent economic risks associated with supply disruptions have originated from the refining and distribution sectors rather than crude oil supplies.

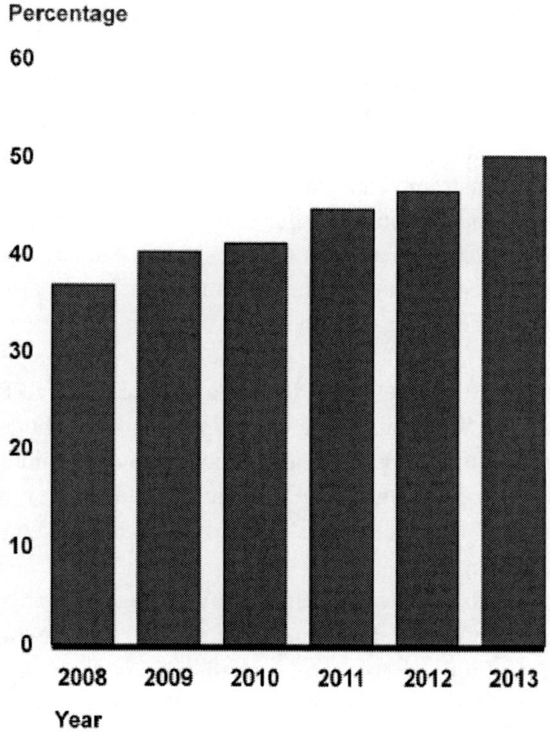

Source: GAO analysis of Energy Information Administration data. | GAO-14-807.

Note: The weight of a crude oil is specified using the American Petroleum Institute (API) gravity standard, which measures the weight of crude oil in relation to water. Data in this figure are imports of crude oils with 25 or lower API gravity as a share of total crude oil imports.

Figure 4. Heavier Crude Oil as a Percentage of Total U.S. Crude Oil Imports, 2008-2013.

DOE Has Taken Steps to Reexamine Some Aspects of the SPR but Has Not Recently Reexamined Its Size

DOE has taken some steps to assess the appropriate location and composition of the SPR in view of changing market conditions, but has not recently re-examined its size. We previously found that federal programs should be re-examined if there have been significant changes in the country or the world that relate to the reason for initiating the program.[52] In that report, we identified a set of reexamination criteria that, when taken together, illustrate the issues that can be addressed through a systematic reexamination process.

We found that many federal programs and policies were designed decades ago to respond to trends and challenges that existed at the time of their creation.

Given fiscal constraints that we are likely to face for years to come, reexamination may be essential to addressing newly emergent needs without unduly burdening future generations of taxpayers. DOE has taken some steps to reexamine how recent changing market conditions could affect the location and composition of the SPR as follows:

- In March 2014, DOE conducted a test sale of SPR crude oil to evaluate the SPR's ability to draw down and distribute SPR crude oil through multiple pipeline and terminal delivery points within one of its distribution systems.[53] DOE officials told us they were reviewing the results of the test sale including data on the movement of crude oil through the system.
- DOE officials also told us they are working to establish a Northeast Regional Refined Petroleum Product Reserve in New York Harbor and New England to store refined consumer fuels. Although the northeast reserve will not store crude oil, it will be considered part of the SPR and hold 1 million barrels of gasoline at a cost of $200 million.
- DOE officials told us that they are conducting a regional fuel resiliency study that will provide insights into whether there is a need for additional regional product reserves and, if so, where these reserves should be located and the capacity.[54] We did not assess this effort because the study was ongoing at the time of our review.
- DOE finalized an assessment in 2010 of the compatibility of crude oil stored in the SPR with the U.S. petroleum refining industry. DOE decided against storing heavy crude oil in the SPR at the time, but committed to revisiting the option of storing heavy crude oil in the future.[55]

However, DOE has not recently reexamined the appropriate size of the SPR. DOE last issued a strategic plan for the SPR in May 2004. The plan outlined the mission, goals, and near-term and long-term objectives for the SPR.[56] In 2006, we recommended that the Secretary of Energy reexamine the appropriate size of the SPR. In 2007, while DOE was planning to expand the SPR to its authorized size of 1 billion barrels, the Administration reevaluated the need for an SPR expansion and decided that the current level was adequate. In responding to our recommendation, DOE stated that its reexamination had taken the form of more "actionable items," including not requesting expansion-funding in its 2011 budget and canceling and redirecting prior year's expansion funding to general operations of the SPR. Officials from DOE's Office of Petroleum Reserves told us that the last time they conducted a comprehensive re-examination of the size of the SPR was in 2005. At that time, DOE's comprehensive study examined the costs and benefits of alternative SPR sizes.[57]

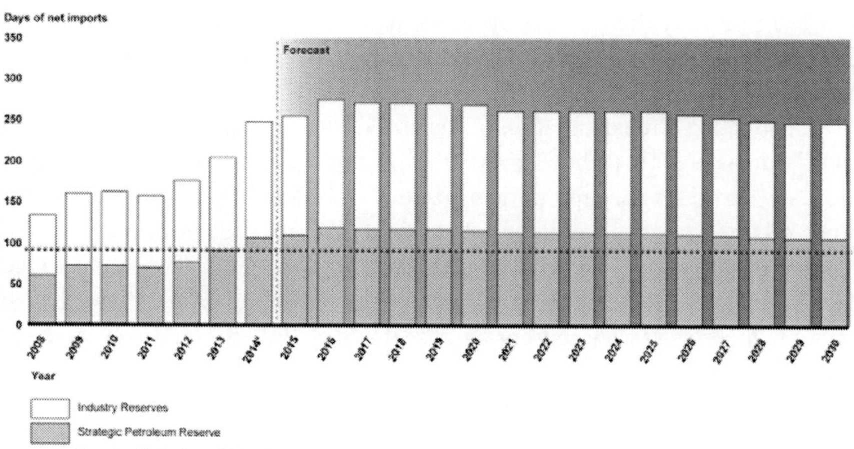

Source: GAO analysis of International Energy Agency data and Energy Information Administration forecasts. | GAO-14-807.

Note: Data for 2015 and later are based on May 2014 reserve levels reported by the International Energy Agency and forecast changes in net imports from the Energy Information Administration's (EIA) reference case forecast. EIA's forecast includes several cases, highlighting uncertainty about future conditions which are not depicted in this figure.

[a]As of May 2014.

Figure 5. United States' Historic and Estimated Compliance with International Energy Agency Obligation to Hold Reserves.

Officials told us that they have not conducted a comprehensive reexamination since 2005 because the SPR only recently met the IEA requirement to maintain 90 days of imports.

However, the IEA requirement is for total reserves, including those held by the government and private reserves. As shown in figure 5, such reserves in the United States are currently in excess of the nation's international obligations and, in some scenarios, are expected to be in excess in the future.

In July 2014, DOE's Office of Inspector General recommended that the Office of Fossil Energy perform a long-range strategic review of the SPR to ensure it is best configured to respond to the current and future needs of the United States.[58]

DOE concurred with the recommendation. DOE stated that it expected to determine the appropriate course of action by August 2014, and according to DOE, it has initiated a process to conduct such a review. The SPR currently holds oil valued at over $73 billion, and without a current reexamination of the SPR's size, DOE cannot be assured that the SPR is sized appropriately.

The SPR may therefore be at risk of holding excess crude oil. In addition, DOE officials told us that SPR infrastructure is aging and will need to be replaced soon.

Conducting a reexamination of the size of the SPR could also help inform DOE's decisions about how or whether to replace existing infrastructure. If DOE were to assess the appropriate size of the SPR and find that it held excess crude oil, the excess oil could be sold to fund other national priorities.

For example, in 1996, SPR crude oil was sold to reduce the federal budget deficit and offset other appropriations. If, for example, DOE found that 90 days of imports was an appropriate size for the SPR, it could sell crude oil worth about $10 billion.[59]

CONCLUSION

Increasing domestic crude oil production, and declines in consumption and crude oil imports have profoundly affected U.S. crude oil markets over the last decade. These changes can have important implications for national energy policies and programs.

The SPR is a significant national asset, and it is important for federal agencies tasked with overseeing such assets to examine how, if at all, changing conditions affect their programs.

DOE has recently taken several steps to reexamine various aspects of the SPR in light of these changes, including its location and composition; however, DOE's most recent comprehensive examination of the appropriate size of the SPR was conducted in 2005 when the general expectation was that the country would increasingly rely on foreign crude oil.

At about that time, however, it began to become clear that this was not to be the case. Removing export restrictions would be expected to lead to further decreases in net imports that would further affect the role of the SPR. Without a reexamination of the SPR that considers whether a smaller or larger SPR is in the national interest in light of current and expected future changes in market conditions, DOE cannot be assured that the SPR is holding an appropriate amount of crude oil in the SPR, and its ability to make appropriate decisions regarding maintenance of the SPR could be compromised.

Recommendation for Executive Action

In view of recent changes in market conditions and in tandem with DOE's ongoing activities to assess the content, connectivity, and other aspects of the SPR, we recommend that the Secretary of Energy undertake a comprehensive reexamination of the appropriate size of the SPR in light of current and expected future market conditions.

Agency Comments and Our Evaluation

We provided a draft of this report to DOE and Commerce for their review and comment.

The agencies provided technical comments, which we incorporated as appropriate. In its written comments, DOE concurred in principle with our recommendation. However, DOE stated that conducting a study of only the size of the SPR would be too narrow in scope and would not address other issues relevant to the SPR carrying out its mission of providing energy security to the United States. DOE stated that a broader, long-range review of the SPR is needed.

We agree that such a review would be beneficial. We do not recommend that DOE undertake an isolated reexamination of the size of the SPR, but that such a reexamination be conducted in tandem with DOE's other activities to assess the SPR and we clarified our recommendation accordingly.

Sincerely yours,
Frank Rusco
Director,
Natural Resources and Environment

APPENDIX I: ADDITIONAL INFORMATION ON FOUR STUDIES OF THE IMPLICATIONS OF REMOVING CRUDE OIL EXPORT RESTRICTIONS

We identified four studies that examined the price and other implications of removing crude oil export restrictions. These four studies are as follows:

- Resources for the Future (RFF). Crude Behavior: How Lifting the Export Ban Reduces Gasoline Prices in the United States. Washington, D.C.: Resources for the Future, February 2014, revised March 2014.
- ICF International and EnSys Energy (ICF International). The Impacts of U.S. Crude Oil Exports on Domestic Crude Production, GDP, Employment, Trade, and Consumer Costs. Washington, D.C.: ICF Resources, March 31, 2014.
- IHS. US Crude Oil Export Decision: Assessing the impact of the export ban and free trade on the US economy. Englewood, Colorado: IHS, 2014.
- NERA Economic Consulting (NERA). *Economic Benefits of Lifting the Crude Oil Export Ban.* Washington, D.C.: NERA Economic Consulting, September 9, 2014.

Table 3 describes these studies and several key assumptions, Table 4 summarizes their findings regarding prices, and Table 5 summarizes their findings regarding other implications of removing crude oil export restrictions.

Table 3. Description of Approach and Key Assumptions in Four Studies on the Implications of Removing Crude Oil Export Restrictions

	Resources for the Future	**ICF International**	**IHS**	**NERA**
Sponsor	Resources for the Future	American Petroleum Institute	Several oil companies.	The Brookings Institution
Description of approach	Using a small static simulation model calibrated to world oil market conditions for 2012, assessed long-run implications of removing export restrictions assuming crude oil prices change to reflect only crude oil qualities.	Used a detailed integrated U.S. and global refining and logistics model together with a detailed assessment of crude oil production and supply and demand changes. Modeled a complete lifting of export restrictions in 2015 and two cases with different assumptions about how hard it will be to accommodate increases in crude oil production.[a]	A bottom up study of crude oil production, refining, and international markets. Examined two outlooks for U.S. crude oil production—a base case based on known production areas and limited technological improvements; and a more optimistic case that includes additional production areas and technological improvements.	Used a partial equilibrium model of the petroleum industry to estimate crude oil production, refining, consumption, and trade effects; and a computable general equilibrium model of the U.S. economy to assess economic effects. Modeled eighteen cases including a reference case and a high oil and gas resources case[b]
Time period covered in analysis	Modeled long run adjustment, not specific years.	2015-2035	2016-2030	2015-2035
Detailed modeling of crude oil production?	No	Yes	Yes	No

Table 3. (Continued)

	Resources for the Future	ICF International	IHS	NERA
Detailed modeling of refineries?	No	Yes	Yes	Yes
Transportation network model?	No	Yes	Yes	Yes
Crude oil production response from OPEC and other producers	Assumed long run world crude oil supply elasticity of 0.4 based on literature. No significant OPEC response modeled.	Assumed world supply response derived from EIA Annual Energy Outlook sensitivity cases. OPEC production would decline by 50,000 barrels per day from 2015 through 2035 without export restrictions.	We were unable to determine IHS' assumption regarding the crude oil production response of other nations.	Examined three cases of OPEC responses: 1) OPEC competes in the market; 2) OPEC maintains crude oil exports and 3) OPEC cutscrude oil exports to maintain crude oil price.

Sources: GAO analysis of Resources for the Future, ICF International, IHS, and NERA studies | GAO-14-807.

[a]ICF assessed two cases; a low case that assumed relatively rapid accommodation of light crude oil, and a high case that assumed slower adaptations to changing crude oil production than were assumed in the low case. In the high case, the resulting differences in prices between domestic and international crude oils would remain wide for a longer period of time.

[b]NERA's reference case assumes U.S. crude oil production peaks in the early part of the next decade and declines thereafter, while the "high oil and gas resources" case represents a more optimistic and sustained view of future production. NERA examined a total of 18 cases with alternative assumptions about (1) future U.S. crude oil production, (2) the international crude oil market, (3) the status, timing, and scope of export restrictions; and (4) OPEC's response to changes in U.S. export restrictions.

Table 4. Summary of Price Implications of Removing Crude Oil Export Restrictions in Four Studies

	Resources for the Future	ICF International	IHS	NERA[a]
U.S. Crude Oil Price	Midwest refiner acquisition costs increase $6.68 per barrel.	West Texas Intermediate prices increase $2.35 to $4.19 per barrel on average.	$7.89 per barrel increase on average.	Prices increase $1.74 per barrel in the reference case and $5.95 per barrel in the high case, on average.
International Crude Oil Prices	Prices increase outside the Midwest by $0.15 per barrel.	Price of Brent declines by $0.37 to $0.79 per barrel on average.	Brent average price declines by $3.24-5.41.	International average crude oil prices decline by $1.31 per barrel in the reference case and $6.23 per barrel in the high case.
U.S. Consumer Fuel Prices	Gasoline prices decline by 1.8 to 4.6 cents per gallon on average.	Petroleum product prices decline by 1.5 to 2.4 cents per gallon on average.	Gasoline prices decline by 9 to 13 cents per gallon on average.	Petroleum product prices decline by 3 cents per gallon in the reference case and 11 cents per gallon in the high case. Gasoline prices decline by 3 cents per gallon in the reference case and 10 cents per gallon in the high case.

Sources: GAO analysis of Resources for the Future, ICF International, IHS, and NERA studies | GAO-14-807.

Note: Price implications in this table are for the period covered in each study. For Resources for the Future, this is the long run price implication; for ICF International and NERA, it is the average from 2015 through 2035; and for IHS, it is the average from 2016 through 2030. Estimates are in 2014 year dollars.

[a]Results refer to the difference between the reference case and its baseline with export restrictions in place, and the difference between the high oil and gas recovery case and its corresponding baseline. NERA also found that removing crude oil export restrictions has no measurable effect in the world oil low price case.

Table 5. Summary of Other Implications of Removing Crude Oil Export Restrictions in Four Studies

	Resources For the Future	**ICF International**	**IHS**	**NERA**[a]
Domestic Crude Oil Production	Canada and the Midwest: increase by 84,000 barrels per day (bpd). Rest of the World: increase by 54,000 bpd. Total world production will increase by 138,000 bpd.	Increases 130,000 bpd on average in the low scenario (10.6 to 10.7 million bpd) and 300,000 bpd in the high scenario (10.4 to 10.7 million bpd).	Increase of 1.2 million bpd in the low case (9.5 to 10.7 million bpd) to 2.3 million bpd in the high case (11 to 13.3 million bpd).	Increase of 0.7 million bpd, on average, in the reference case and 3.3 million bpd in the high case.
Oil Industry Investments	Not addressed.	Additional $16 to $73 billion in additional crude oil production related investment between 2015 and 2020, and a decline in refining investment of $5 to $7 billion from 2015-2035.	Net cumulative investment increases by $806 billion to $1.1 trillion from 2016-2030.	Increases by $1.08 billionper year on average in the reference case and $26.81 billion per year in the high case from 2015 through 2035.[b]
Environmental Implications	Carbon dioxide emissions would increase by 22 million metric tons per year.	Not addressed.	Not addressed.	Greenhouse gas emissions increase by 12million metric tons of carbon dioxide equivalents per year.

	Resources For the Future	ICF International	IHS	NERA[a]
Trade Implications	Would improve trade balance, though not estimated.	Volumes: Decline in net crude oil imports of 26,000-200,000 bpd and decline in net petroleum product imports of nearly 93,000 bpd. Balance of trade: Increases by $8 to $15 billion.	Volumes: Project a decline in net imports, specific amount not provided. Balance of trade: Increases net petroleum trade by $72-$101 billion per year.	Volumes: Decline in net crude oil and petroleum product imports of 0.63 million bpd in the reference case and 3.15 million bpd in the high case.

Sources: GAO analysis of Resources for the Future, ICF International, IHS, and NERA studies. | GAO-14-807

Notes: Implications in this table are for the period covered in each study. For Resources for the Future, this is in the long run; for ICF International and NERA, it is the average from 2015 through 2035; and for IHS, it is the average from 2016 through 2030. Estimates are in 2014 year dollars.

[a]Results refer to the difference between the reference case and its baseline with export restrictions in place, and the difference between the high oil and gas recovery case and its corresponding baseline. NERA also found that removing crude oil export restrictions has no measurable effect in the world oil low price case.

[b]Includes investment in oil extraction, industrial, and manufacturing sectors.

APPENDIX II: LIST OF STAKEHOLDERS INTERVIEWED

This appendix lists the stakeholders we interviewed. Stakeholders included representatives of companies and interest groups with a stake in the outcome of decisions regarding crude oil export restrictions, as well as academic, industry, and other experts.

1) American Fuel & Petrochemical Manufacturers
2) American Petroleum Institute
3) American Automobile Association
4) Jason Bordoff, Columbia School of International and Public Affairs
5) Severin Borenstein, University of California, Berkeley
6) Stephen Brown, Resources for the Future and University of Nevada-Las Vegas
7) Citigroup
8) Deborah Gordon, Carnegie Endowment for International Peace
9) Hillard Huntington, Stanford University
10) IHS
11) ICF International and EnSys Energy
12) Kenneth Medlock, Rice University
13) NERA Economic Consulting
14) Lorne Stockman, Oil Change International
15) Philip Verleger, PKVerleger LLC
16) Dan Weiss, Center for American Progress
17) Valero Energy Corporation

End Notes

[1] EIA is a statistical agency within the Department of Energy that collects, analyzes, and disseminates independent information on energy issues.

[2] Proposed legislation includes H. R. 4349, the Crude Oil Export Act, which was referred to the House Foreign Affairs Committee, Subcommittee on Terrorism, Nonproliferation, and Trade on June 10, 2014.

[3] Specifically, we searched sources including Proquest, PolicyFile, and Web of Science in April 2014.

[4] The four studies are: Resources for the Future (RFF), *Crude Behavior: How Lifting the Export Ban Reduces Gasoline Prices in the United States* (Washington, D.C.: Resources for the Future, February 2014, revised March 2014); ICF International and EnSys Energy (ICF International), *The Impacts of U.S. Crude Oil Exports on Domestic Crude Production, GDP, Employment, Trade, and Consumer Costs* (Washington, D.C.: ICF Resources, March

31, 2014); IHS, *US Crude Oil Export Decision: Assessing the impact of the export ban and free trade on the US economy* (Englewood, CO: IHS, 2014); and NERA Economic Consulting (NERA), *Economic Benefits of Lifting the Crude Oil Export Ban* (Washington, D.C.: NERA Economic Consulting, September 9, 2014).

[5] Bloomberg is a provider of business and financial news, data, and analytics.

[6] 15 C.F.R. §754.2(a).

[7] 15 C.F.R. §754.2(b)(2).

[8] For example, the Mineral Leasing Act of 1920 restricts exports of domestically produced crude oil transported by pipeline over certain rights-of-way (30 U.S.C. §185(u)); the Outer Continental Shelf Lands Act restricts exports of crude oil from the outer continental shelf (29 U.S.C. §1354); the Naval Petroleum Reserves Production Act restricts the export of crude oil produced from the Naval Petroleum Reserves (10 U.S.C. §7430) and Section 201 of Pub. L. No. 104-58, "Exports of Alaskan North Slope Oil," provides for exports of domestically produced crude oil transported by pipeline over rights-of-way granted pursuant to section 203 of the Trans-Alaska Pipeline Authorization Act (30 U.S.C. §185(s)).

[9] 15 C.F.R. §754.2(c).

[10] NERA, *Economic Benefits of Lifting the Crude Oil Export Ban*, p.20.

[11] The 29 member countries of the IEA are Australia, Austria, Belgium, Canada, Czech Republic, Denmark, Estonia, Finland, France, Germany, Greece, Hungary, Ireland, Italy, Japan, Republic of Korea, Luxembourg, the Netherlands, New Zealand, Norway, Poland, Portugal, Slovak Republic, Spain, Sweden, Switzerland, Turkey, the United Kingdom, and the United States.

[12] Under the agreement, capacity to switch to nonpetroleum-based fuels and standby crude oil production capacity can also be used to meet the reserve requirement.

[13] Pub. L. No. 94-163, §161, 89 Stat. 888-889 (1975), codified as amended at 42 U.S.C. §6241. The statute provides for a drawdown of the reserve upon a finding by the President that there is a "severe energy supply interruption" as well as an event that is, or is likely to become, an energy supply shortage "of significant scope or duration" (42 U.S.C. §6241(d), (h)).

[14] For more information on these technologies, see: GAO, *Oil and Gas: Information on Shale Resources, Development, and Environmental and Public Health Risks,* GAO-12-732 (Washington, D.C.: Sept. 5, 2012).

[15] GAO, *Petroleum Refining: Industry's Outlook Depends on Market Changes and Key Environmental Regulations,* GAO-14-249 (Washington, D.C.: Mar. 14, 2014).

[16] Because of the large number of grades of crude oils, buyers and sellers use benchmark crude oils as a reference in pricing crude oil. A benchmark crude oil is typically an abundantly produced and frequently traded crude oil. For example, crude oils produced in North and South America are typically priced in reference to WTI.

[17] The density, or gravity of a crude oil is specified using the American Petroleum Institute (API) gravity standard, which measures the weight of crude oil in relation to water, which has an API gravity of 10 degrees. For the purposes of this estimate, we considered light oils as those with an API gravity of 35 degrees or above. See: Energy Information Administration, *U.S. Crude Oil Production Forecast-Analysis of Crude Types* (Washington, D.C.: May 29, 2014).

[18] In general, heavier crude oils require more complex and expensive refineries to process the crude oil into usable products, but have been less expensive to purchase than lighter crude oils.

[19] Stakeholders we interviewed include representatives of companies and interest groups with a stake in the outcome of decisions regarding crude oil export restrictions, as well as academic, industry, and other experts.

[20] Increasing U.S. crude oil production may also have affected some global oil prices. For example, in 2013 U.S. crude oil production grew more than the combined increase in the rest of the world, which contributed to relatively stable global crude oil prices in 2013, according to EIA. See: EIA, *Today in Energy: U.S. Crude Oil Production Growth Contributes to Global Oil Price Stability in 2013* (Washington, D.C.: Jan. 9, 2014).

[21] We summarize the results of four studies: Resources for the Future (RFF), *Crude Behavior: How Lifting the Export Ban Reduces Gasoline Prices in the United States* (Washington, D.C.: Resources for the Future, February 2014, revised March 2014); ICF International and EnSys Energy (ICF International), *The Impacts of U.S. Crude Oil Exports on Domestic Crude Production, GDP, Employment, Trade, and Consumer Costs* (Washington, D.C.: ICF Resources, March 31, 2014); IHS, *US Crude Oil Export Decision: Assessing the impact of the export ban and free trade on the US economy* (Englewood, CO: IHS, 2014); and NERA Economic Consulting (NERA), *Economic Benefits of Lifting the Crude Oil Export Ban* (Washington, D.C.: NERA Economic Consulting, September 9, 2014). For additional information on these studies, see appendix I.

[22] Specifically, we reported on three forecasts, from the International Energy Agency (IEA), IHS Global, and EIA. See, GAO-14-249.

[23] For example, EIA estimates that light crude oil production may increase by 513,000 barrels per day in 2015. (See: EIA, *U.S. Crude Oil production Forecast-Analysis of Crude Types* (Washington, D.C.: May 29, 2014).

[24] Specifically, the Department of Commerce's definition of crude oil includes condensates, which are light liquid hydrocarbons recovered primarily from natural gas wells.

[25] Specifically, companies often process condensate through stabilization units to reduce their volatility and prepare the condensate for transport to markets. Some stabilization units include distillation towers. In March and May 2014, the Department of Commerce issued commodity classifications that determined that condensates processed through a crude oil distillation tower, as described by the two companies requesting clarification, did not meet the definition of crude oil in BIS's regulations and thus were not subject to the export prohibitions applicable to U.S. produced crude oil.

[26] This clarification provided by the Department of Commerce occurred after the publication of the RFF, ICF International, and IHS studies and thus this was not taken into consideration in the studies. NERA also did not consider the potential effect of the clarification in its study.

[27] Unless otherwise noted, dollar estimates in the rest of this report have been converted to 2014 year dollars. These are average price effects over the study time frames, and some cases in some studies project larger price effects in the near term that decline over time.

[28] RFF also estimates a decrease in consumer fuel prices but this decrease is as a result of increased refinery efficiency (even with an estimated slight increase in the international crude oil price).

[29] See Severin Borenstein and Ryan Kellogg, "The Incidence of an Oil Glut: Who Benefits from Cheap Crude Oil in the Midwest?" *The Energy Journal 35, no. 1* (2014).

[30] The Merchant Marine Act of 1920, also known as the Jones Act, in general, requires that any vessel (including barges) operating between two U.S. ports be U.S.-built, -owned, and -operated.

[31] EIA, *Potential Impacts of Reductions in Refinery Activity on Northeast Petroleum Product Markets* (Washington, D.C.: May 11, 2012).

[32] U.S. crude oil production grows from 2012 through 2019 in EIA's Reference scenario, peaking at more than 9.6 million barrels per day—about 3.1 million barrels per day above the 2012 total and close to the historical high of 9.6 million barrels per day in 1970. However, EIA projects declines in later years. EIA examined several alternative scenarios, including a High Oil and Gas Resource scenario, where growth in oil production continues for a longer period of time than projected in the Reference scenario. Domestic crude oil production increases to nearly 13 million barrels per day before 2035 in this scenario. EIA also examined a Low Oil and Gas Resource scenario reflecting uncertainty about tight oil and shale crude oil and natural gas resources, leading to lower domestic production than in the Reference scenario. In the latter, production reaches 9.1 million barrels per day in 2017 before falling to 6.6 million barrels per day in 2040 (Energy Information Administration. *Annual Energy Outlook 2014*. May 7, 2014. See: http://www.eia.gov/forecasts /aeo/MT_liquidfuels.cfm.)

[33] In addition, RFF estimated that oil production in Canada and in the Midwest United States would gradually increase if the restrictions were lifted by about 84,000 barrels per day. RFF estimated production elsewhere in the United States and the rest of the world would increase by 54,000 barrels per day for a total increase in world production of 138,000 additional barrels per day. IHS projected an additional 1.2 to 2.3 million barrels per day of crude oil production from 2016- through 2030. (See app. I for additional information.)

[34] U.S. Department of Energy. Energy Information Administration. Country Report: India. See: http://www.eia.gov/countries/country-data.cfm?fips=IN. June 26, 2014.

[35] GAO-12-732.

[36] We found, for example, that water is needed for five distinct groups of activities that occur during the life cycle of oil shale development: (1) extraction and retorting, (2) upgrading of shale oil, (3) reclamation, (4) power generation, and (5) population growth associated with oil shale development. We reviewed a set of studies that indicated that the expected total water needs for the entire life cycle of oil shale production ranged from about 1 barrel (or 42 gallons) to 12 barrels of water per barrel of oil produced from in situ (underground heating) operations, with an average of about 5 barrels, and from about 2 to 4 barrels of water per barrel of oil produced from mining operations with surface heating. See GAO, *Energy-Water Nexus: A Better and Coordinated Understanding of Water Resources Could Help Mitigate the Impacts of Potential Oil Shale Development,* GAO-11-35 (Washington, D.C.: Oct. 29, 2010).

[37] GAO-12-732.

[38] Burning natural gas is known as flaring, while releasing natural gas directly into the atmosphere is called venting. In 2004, we found that venting and flaring have adverse environmental effects and result in loss of a significant amount of energy. In areas where the primary purpose of drilling is to produce oil, producers flare or vent because no local market exists for the gas. See GAO, *Natural Gas Flaring and Venting: Opportunities to Improve Data and Reduce Emissions,* GAO-04-809 (Washington, D.C.: July 14, 2004).

[39] Ceres, *Flaring Up: North Dakota Natural Gas Flaring More Than Doubles in Two Years* (July 2013). Ceres is a nonprofit organization.

[40] According to the Environmental Protection Agency, carbon dioxide is the primary greenhouse gas emitted through human activities. Further, the main human activity that emits carbon dioxide is the combustion of fossil fuels (coal, natural gas, and oil) for energy and transportation.

[41] Carbon dioxide equivalents provide a common standard for measuring the warming potential of different greenhouse gases and are calculated by multiplying the emissions of the non-carbon dioxide gas by its global warming potential, a factor that measures its heat-trapping ability relative to that of carbon dioxide. The NERA study suggests that maintaining crude oil export restrictions could be a relatively costly means of constraining greenhouse gas emissions. The authors calculated that the cost per ton of avoided emissions using crude oil export restrictions range from about $1,200 to $1,400 per ton of carbon dioxide in one case, and about $900 to $1,000 per ton in another case (in 2013 dollars). The authors stated that these costs are higher than government estimates of the benefits of reducing carbon dioxide emissions—in 2013, an interagency working group estimated the benefits of avoided emissions ranged from $37 to $56 per metric ton of carbon dioxide from 2015 through 2035 (in 2007 dollars at a 3 percent discount rate). See: Interagency Working Group on the Social Cost of Carbon, United States Government, *Technical Support Document: Technical Update of the Social Cost of Carbon for Regulatory Impact Analysis – Under Executive Order 12866* (Washington, D.C.: May 2013, revised November, 2013).

[42] GAO-14-249.

[43] Department of Transportation data show that the majority of the 2013 incidents were small; however, two incidents in 2013, in Aliceville, Alabama, and Casselton, North Dakota, resulted in large spills and greater damage. Significant incidents have continued to occur in 2014, including an April derailment and fire in Lynchburg, Virginia. The Department of Transportation Pipeline and Hazardous Materials Safety Administration's incident database can be searched at https://hazmatonline.phmsa.dot.gov/IncidentReportsSearch/ (accessed June 12, 2014). GAO, *Oil and Gas Transportation: Department of Transportation Is Taking Actions to Address Rail Safety, but Additional Actions are Needed to Improve Pipeline Safety*, GAO-14-667 (Washington, D.C.: August 21, 2014).

[44] Growth in one sector of the economy can result in economy-wide growth through follow-on effects. For example, researchers at the Federal Reserve Bank of Dallas found that oil development in the Eagle Ford region of South Texas has had profound effects on jobs, income, and spending in the region with effects beyond those in the oil sector alone. See: Gilmer, Robert W., Raúl Hernandez, and Keith Phillips, "Oil Boom in Eagle Ford Shale Brings New Wealth to South Texas," *Southwest Economy* (Federal Reserve Bank of Dallas: Second Quarter, 2012).

[45] According to NERA, because of the increase in economic growth triggered by investment in more production capacity and infrastructure, there will be a corresponding acceleration of the rate at which the economy moves toward full employment.

[46] President's Council of Economic Advisers, *The All-of-the-Above Energy Strategy as a Path to Sustainable Economic Growth*. Executive Office of the President. May 2014.

[47] According to IEA, all reserves held by industry count towards meeting a country's IEA reserve commitment. Most member governments require certain companies, such as importers, refiners, product suppliers or wholesalers, to hold a minimum number of days of reserves. However, 9 countries, including the United States, do not place such a requirement on industry. According to DOE officials, commercial entities set their reserves based on the economic principle of inventory minimization and require a large portion of the reserves they hold to keep their logistical systems moving. These reserves may therefore not be available to address a supply disruption in the same way as mandated industry reserves or public reserves.

[48] DOE officials cited several studies to show the diversity of views in this regard, and noted that most of the literature on the effect of supply disruptions on the economy does not focus on

net imports, and that this is an area of active research. According to DOE, the literature shows that the United States is still vulnerable to oil price shocks as these can cause dislocations within key economic sectors. As part of this review, we did not assess the validity of the studies. For example, see: J.D. Hamilton, "Causes and Consequences of the Oil Shock of 2007-08," *Brookings Papers on Economic Activity*, Spring 2009: 215-261; L. Kilian, "The Economic Effects of Energy Price Shocks," *Journal of Economic Literature* 46, no.4(2008): 871-909.

[49] GAO, *Strategic Petroleum Reserve: Available Oil Can Provide Significant Benefits, but Many Factors Should Influence Future Decisions about Fill, Use, and Expansion,* GAO-06-872 (Washington, D.C.: Aug. 24, 2006).

[50] For example, according to DOE, storing heavier crude oil in the SPR would require upgrading the Sun Terminal, a tanker delivery location, to handle the heavier crude oil; building an additional pipeline from one of four storage sites to maintain current drawdown rates; and performing other site improvements. U.S. Department of Energy. Office of the Deputy Assistant Secretary for Petroleum Reserves. *Strategic Petroleum Reserve: Updated Crude Compatibility Study*. April 2010.

[51] DOE also operates the Northeast Home Heating Oil Reserve, which stores 1 million barrels of diesel fuel in Massachusetts and Connecticut.

[52] *21st Century Challenges: Reexamining the Base of the Federal Government,* GAO-05-325SP (Washington, D.C.: Feb. 1, 2005).

[53] According to DOE, the SPR's oil storage facilities are grouped into three geographical distribution systems in the Gulf Coast: Seaway, Texoma, and Capline. Each system has access to one or more major refining centers, interstate crude oil pipelines, and marine terminals for crude oil distribution.

[54] This study is being undertaken as part of DOE's Quadrennial Energy Review. According to DOE, the Quadrennial Energy Review will provide a multiyear road map that outlines federal energy policy objectives, legislative proposals to Congress, executive actions, and resource requirements. The first installment of the review will focus on transmission, storage, and distribution infrastructure that links energy supplies, including crude oil, to intermediate users.

[55] U.S. Department of Energy. Office of the Deputy Assistant Secretary for Petroleum Reserves. *Strategic Petroleum Reserve: Updated Crude Compatibility Study*. April 2010.

[56] U.S. Department of Energy. Office of Fossil Energy. *Strategic Petroleum Reserve (SPR) Strategic Plan*. May 19, 2004.

[57] The report evaluates the net economic benefits of enhancing the SPR size and drawdown capability. DOE assessed alternative SPR sizes and drawdown capabilities using a numerical simulation model, considering the benefits and costs of oil stockpiling. The evaluation included the SPR's ability to reduce economic losses and oil import costs during oil shocks, and it subtracted the costs of building, filling, and operating the reserve. See: Paul Leiby and David Bowman. *Economic Benefits of Expanded Strategic Petroleum Reserve Size or Drawdown Capability*. Final Report, Oak Ridge National Laboratory. ORNL/TM-2006/5. Dec. 31, 2005.

[58] According to DOE officials, this review should take into consideration what the near-term and long-term role of the reserve should be relative to U.S. energy and economic security goals and objectives and International Energy Program requirements; what the optimal configuration and capabilities (e.g., composition, volume, location of petroleum products, infrastructure requirements, distribution capability, and performance criteria) of the Reserve should be; the resources required to attain and maintain the Reserve's long-term

sustainability (to ensure alignment with optimal configuration and capabilities); and whether existing legal authorities that govern the policies, configuration, and capabilities of the Reserve are adequate to ensure the Reserve can meet both current and future U.S. energy and economic security goals and objectives. U.S. Department of Energy. Office of Inspector General. Office of Audits and Inspections. *The Strategic Petroleum Reserve's Drawdown Readiness.* DOE/IG-0916. July 2014.

[59] Calculated based on DOE's assessment of the weighted average price of crude oil held in the SPR as of May 28, 2014.

In: U.S. Exports of Oil and Natural Gas
Editor: Frank G. Adams
ISBN: 978-1-63463-904-0

Chapter 2

NATURAL GAS: FEDERAL APPROVAL PROCESS FOR LIQUEFIED NATURAL GAS EXPORTS*

United States Government Accountability Office

ABBREVIATIONS

CAA	Clean Air Act
Coast Guard	U.S. Coast Guard
Corps	U.S. Army Corps of Engineers
DOE	Department of Energy
DOT	Department of Transportation
EA	environmental assessment
EIA	Energy Information Administration
EIS	environmental impact statement
EPA	Environmental Protection Agency
FERC	Federal Energy Regulatory Commission
FTA	free trade agreement
LNG	liquefied natural gas
MARAD	Maritime Administration
MMBtu	million British thermal unit

* This is an edited, reformatted and augmented version of The United States Government Accountability Office publication, GAO-14-762, dated September 2014.

NEPA	National Environmental Policy Act
NERA	NERA Economic Consulting
NGA	Natural Gas Act of 1938
NSR	new source review
PHMSA	Pipeline and Hazardous Materials Safety Administration

WHY GAO DID THIS STUDY

Technological advances in hydraulic fracturing and horizontal drilling have resulted in a dramatic increase in the amount of natural gas that can be produced domestically. DOE is responsible for reviewing applications to export LNG—natural gas cooled to a liquid state for transport—and, under the Natural Gas Act, must approve an application unless it finds that approval is not consistent with the public interest. Since 2010, DOE has received 35 applications to export LNG that must address the public interest question. In addition, under NEPA, FERC is required to assess how LNG export facilities may affect the environment and is responsible for granting approval to build and operate export facilities. Since 2010, FERC has received 17 applications to construct export facilities.

GAO was asked to report on the federal process for reviewing applications to export LNG. This report describes (1) the status of applications to export LNG and DOE's process to review them and (2) the status of applications to build LNG export facilities and FERC's process to review them.

GAO reviewed laws, regulations, and guidance; examined export approvals; visited LNG facilities; and interviewed federal and state agency officials and industry representatives, including LNG export permit applicants.

GAO is not making any recommendations in this report.

WHAT GAO FOUND

Since 2010, of 35 applications it has received that require a public interest review, the Department of Energy (DOE) has approved 3 applications to export liquefied natural gas (LNG) and 6 applications are conditionally approved with final approval contingent on the Federal Energy Regulatory

Commission's (FERC) issuance of a satisfactory environmental review of the export facility. DOE considers a range of factors to determine whether each application is in the public interest. After the first application was conditionally approved in 2011, DOE commissioned a study to help it determine whether additional LNG exports were in the public interest. Since the 16-month study was published in December 2012, DOE issued 7 conditional approvals (one of which became final) and 1 other final approval. In August 2014, DOE suspended its practice of issuing conditional approvals; instead, DOE will review applications after FERC completes its environmental review.

Since 2010, FERC has approved 3 LNG export facilities for construction and operation, including 2 facilities in 2014, and is reviewing 14 applications. FERC's review process is, among other things, designed to fulfill its responsibilities under the National Environmental Policy Act (NEPA). Before submitting an application to FERC, applicants must enter an initial stage called pre-filing to identify and resolve potential issues during the earliest stages of a project. Of the 14 applications, 5 are in the pre-filing stage at FERC and not shown in the figure below. FERC conducts an environmental and safety review with input from other federal, state and local agencies.

* * *

September 26, 2014

The Honorable Lisa Murkowski
Ranking Member
Committee on Energy and Natural Resources
United States Senate

Dear Senator Murkowski:

Technological advances in hydraulic fracturing and horizontal drilling have dramatically increased the amount of natural gas that can be extracted domestically. This increase has helped reduce natural gas prices in the United States, although prices remain significantly higher overseas. To take advantage of overseas prices, energy companies have applied to the federal government for permission to export natural gas. Natural gas, like crude oil, requires federal approval before it can be exported.[1] To transport natural gas by ship, the natural gas is cooled to a liquid state called liquefied natural gas (LNG).

This cooling process requires expensive processing and liquefaction facilities that do not currently exist in the contiguous United States.[2]

The Department of Energy (DOE) and Federal Energy Regulatory Commission (FERC) are two federal agencies with primary responsibility for approving LNG exports. Under the Natural Gas Act of 1938 (NGA), as amended,[3] DOE is responsible for reviewing LNG export applications and, for countries that do not have a free trade agreement (FTA) with the United States, determining whether approval of such applications is consistent with the public interest.[4] In this report, we discuss non-FTA applications that require a public interest review. FERC is responsible for authorizing the construction and operation of facilities located onshore or within state waters that liquefy natural gas and load the LNG onto ships for export.[5] FERC conducts an environmental review of the facility to help with this decision.[6] Until May 2014, DOE granted approvals—known as conditional approvals—on the condition that FERC completed a satisfactory environmental review of the associated export facility. DOE granted final approvals only after it had the opportunity to consider and adopt the FERC-led review. In May 2014, DOE proposed suspending its practice of issuing conditional approvals and instead proposed reviewing applications only after FERC has completed the environmental review, and DOE has sufficient information to determine the public interest of an application.[7] In August 2014, DOE finalized this proposal after a public comment and review period.[8] Since 2010, DOE has received 35 applications to export LNG that require a public interest review, and FERC has received 17 applications to construct LNG export facilities.[9]

You asked us to report on DOE's and FERC's processes for reviewing applications for LNG exports and export facilities, respectively. This report describes (1) DOE's process for reviewing applications to export LNG and the status of those applications and (2) FERC's process for reviewing applications to build LNG export facilities and the status of those applications.

To describe the process and status of application reviews at DOE and FERC, we reviewed DOE and FERC approval documents and spoke with agency officials. To describe DOE's process for reviewing applications to export LNG, we examined relevant federal laws, regulations, and policy documents. We reviewed DOE approvals to export LNG, including one final and seven conditional approvals. We also interviewed DOE officials responsible for reviewing applications to export LNG to discuss the review process and how they determine whether an application is in the public interest. To describe FERC's process for reviewing applications to build LNG export facilities, we examined relevant federal laws, regulations and FERC guidance.

We also interviewed FERC officials to discuss the export facility review process. We reviewed FERC environmental documents relating to LNG export facilities. We also interviewed other federal and state agencies involved in the review process—including the U.S. Coast Guard (Coast Guard), the Pipeline and Hazardous Materials Safety Administration (PHMSA) within the Department of Transportation (DOT), the U.S. Army Corps of Engineers (Corps), the Maryland Public Service Commission, the Louisiana Department of Environmental Quality, and the Texas Commission on Environmental Quality—to describe their roles in FERC's review process. To describe the applicants' experience, we interviewed a nonprobability sample of applicants from the Sabine Pass, Freeport, Lake Charles, Cove Point, and Jordan Cove facilities. We also visited a nonprobability sample of LNG facilities at Lusby, Maryland; Brazoria County, Texas; and Cameron Parish, Louisiana.[10] We selected these facilities because they applied to FERC and were approved or conditionally approved for export by DOE. To describe how the public is involved in the review process, we interviewed representatives of organizations that have commented on applications, such as America's Energy Advantage, the Sierra Club, the American Public Gas Association, and the Industrial Energy Consumers of America.

We conducted this performance audit from September 2013 to September 2014 in accordance with generally accepted government auditing standards. Those standards require that we plan and perform the audit to obtain sufficient, appropriate evidence to provide a reasonable basis for our findings and conclusions based on our audit objectives. We believe that the evidence obtained provides a reasonable basis for our findings and conclusions based on our audit objectives.

BACKGROUND

This section describes the factors affecting recent LNG activities in the United States, the liquefaction process, DOE's responsibilities for authorizing export applications, FERC's responsibilities for authorizing export facilities, and positions of those supporting or opposing the export of LNG.

Factors Affecting Recent LNG Activities in the United States

According to the Congressional Research Service, in the early 2000s, natural gas production in the United States was declining as energy demand was increasing and, as recently as the mid-to-late 2000s the United States was projected to be a growing natural gas importer.[11] In addition to four onshore import terminals that were already operational, natural gas companies built five LNG onshore import facilities in the latter half of the 2000s to meet the expected need for natural gas imports.[12] However, technology enhancements improved the extraction of natural gas from shale formations and resulted in a dramatic increase in domestic natural gas production. These technology enhancements allow companies to extract natural gas from shale formations that were previously considered to be inaccessible because traditional techniques did not yield sufficient amounts for economically viable production.[13] According to Energy Information Administration (EIA) data,[14] between 2007 and 2013, domestic natural gas withdrawals increased by 22 percent, driven primarily by increased withdrawals from shale formations.[15] According to EIA, increases in natural gas supplies generally cause prices to drop. Specifically, between 2007 and 2013, the price of natural gas in the United States decreased by nearly 50 percent.[16] As the price of natural gas in the United States declined, prices in Europe and Asia remained considerably higher. In July 2014, FERC estimated that prices of LNG imported to Europe and Asia during August of 2014 would be about 100 and 250 percent higher than prices in the United States, respectively.[17] These price differences have motivated U.S companies to apply to export natural gas.

The Liquefaction Process

The majority of U.S. trade in natural gas is by pipeline with Canada and Mexico; however, over long distances separated by water, natural gas is generally converted to LNG and transported by specialized tanker ship. [18] To convert natural gas to LNG, companies pretreat the natural gas to remove components that would freeze during the liquefaction process and contaminate the LNG.[19] After the gas is pretreated, it is processed through a complex system called a liquefaction train that cools the natural gas to -260 degrees Fahrenheit, converting it to a liquid state. This process reduces the volume of the gas by 600 times. Once liquefied, the natural gas is stored in large tanks until it is offloaded to a ship for transport. Once the ship reaches its destination, it is offloaded to tanks for storage or converted to natural gas for distribution by pipeline.[20] Figure 1 illustrates some of the common components of an LNG export facility.

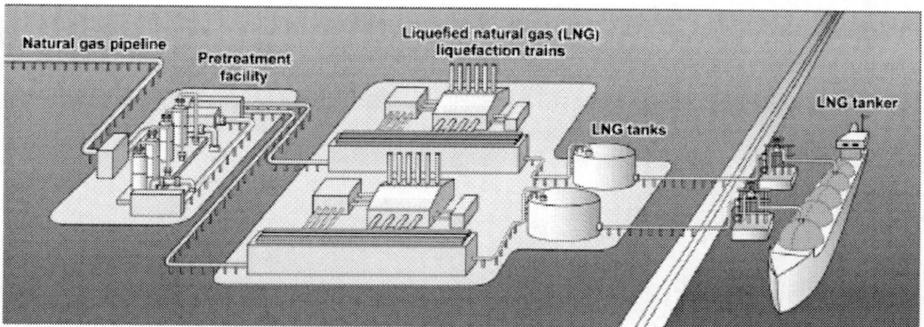

Sources: GAO analysis of induistry documents. GAO-14-762.

Figure 1. LNG Export Facility.

DOE's Responsibilities

Under Section 3 of the NGA, the import or export of LNG and the construction or expansion of LNG facilities requires authorization from DOE.[21] In 1984, DOE delegated the responsibility to approve or deny applications for LNG facilities to FERC.[22] Under Section 3, an authorization is to be granted unless DOE finds that approving the export or import is inconsistent with the public interest. According to DOE, Section 3(a) of the NGA creates a rebuttable presumption that a proposed export of natural gas is in the public interest—that is, it places the burden on those opposing an application to demonstrate that an export is inconsistent with the public interest. The NGA also authorizes DOE to attach terms and conditions necessary to protect the public interest. DOE evaluates public interest under Section 3, and can conduct studies or other reviews to support its public interest determination.

In the Energy Policy Act of 1992, Congress amended the NGA to require DOE to use a different standard for the review of applications for export to countries with FTAs with the United States (FTA countries).[23] Specifically, under Section 3(c) of the NGA, DOE must treat applications to export LNG to FTA countries as consistent with the public interest, and DOE is to approve these applications without modification or delay.[24] These FTA applications therefore do not require the same public interest review as non-FTA applications. DOE started to receive applications to export LNG in 2010 and, since then, it has approved 37 of 42 applications to export LNG to FTA countries. During this same period, DOE approved 9 (3 final and 6 conditional) of 35 applications to export LNG to non-FTA countries.[25] Most major importers of LNG are non-FTA countries such as Japan and India, among others. As previously mentioned, this report discusses DOE's process to review applications to export to non-FTA countries.

FERC's Responsibilities

In keeping with its obligation to authorize LNG facility siting and construction under the NGA, FERC reviews applications to construct and operate LNG export facilities. FERC's review is considered a federal action and subject to the National Environmental Policy Act (NEPA).[26]

NEPA requires federal agencies to assess the projected effects of major federal actions that significantly affect the environment. Prior to the NEPA review, the law requires applicants to communicate with FERC for a minimum of 6 months—known as pre-filing—before submitting an application.[27] FERC acts as the lead agency for the environmental review required by NEPA, prepares the NEPA environmental documentation, and coordinates and sets the schedule for all federal authorizations.[28] The outcome of this review is an environmental document, also called the NEPA document, which provides the commissioners with staff's assessment of the environmental impacts from facility construction and operation.[29]

Positions Surrounding the Export of Natural Gas

DOE and FERC consider comments from the public during the application review process, and these comments reflect a range of perspectives on the potential benefits or harm from exports. Proponents maintain that LNG exports are consistent with U.S. free trade policies and will provide an economic boon for the United States, resulting in increased employment and an improved trade balance among other things.[30] They assert that the increased availability of natural gas resources will prevent a significant increase in natural gas prices. Opponents have expressed numerous environmental and economic concerns about LNG exports. For example, opponents have expressed concern that exports will increase hydraulic fracturing and its associated environmental effects, as well as increase greenhouse gas emissions from the production and consumption of natural gas.[31] Other opponents have expressed concern that exports will increase domestic natural gas prices, hurting the public and the growing industrial and manufacturing sectors that are sensitive to natural gas prices. Opponents have also stated that the primary beneficiaries of LNG exports will be a small segment of society involved in natural gas development and trade, and that most segments of society will lose economically. Evaluating whether exports of LNG to non-FTA countries are consistent with the public interest is beyond the scope of this report.

DOE Has Approved 9 Applications to Export LNG through a Review Process That Considers a Range of Factors to Determine Whether Approval Is in the Public Interest

Since 2010, DOE has granted final approval to 3 applications and conditional approval to 6 others. DOE considers a range of factors to determine whether approving an export application is in the public interest.

DOE Has Approved 3 Applications and Conditionally Approved 6 Others

As of mid-September 2014, DOE has granted 3 final approvals for applications to export LNG, including the Sabine Pass application in 2012 and the Cameron LNG and Carib Energy applications in September 2014. Sabine Pass is the only LNG export facility currently under construction in the United States and is expected to begin operations in late 2015.[32] In August 2011, after DOE conditionally approved exports from Sabine Pass, DOE commissioned a study of the cumulative effects of additional LNG exports on the economy and the public interest.[33] DOE did not approve any conditional applications during the 16-month period of the study.[34] The study was completed in December 2012. Since then, DOE has conditionally approved 7 applications, including the Cameron LNG application that it granted final approval in September 2014 (See fig. 2).[35] DOE also approved the Carib Energy application in September 2014.[36] DOE's export approvals, as of late August 2014, amount to 10.56 billion cubic feet of natural gas per day in the form of LNG; for comparison,

Qatar, the world's largest exporter of LNG, exported about 10 billion cubic feet per day in 2012.

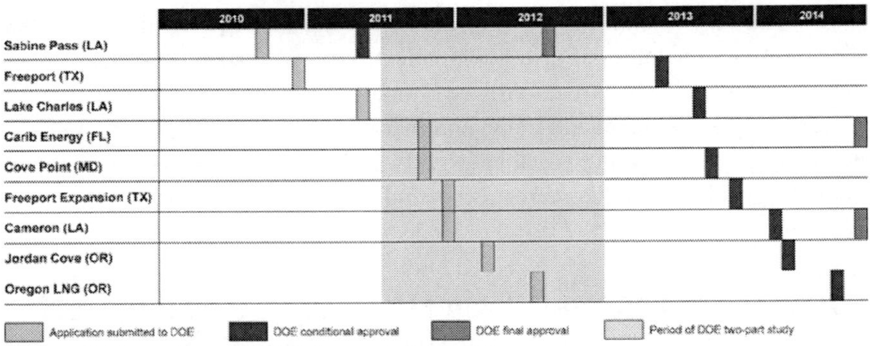

Source: GAO analysis of DOE data. GAO-14-762.

Notes: DOE has received 26 other applications since the beginning of 2010 that DOE has not approved and are not displayed.

The export amount that DOE conditionally approved for the Freeport and Freeport Expansion applications is accounted for in a single Freeport facility application to FERC.

Figure 2. Status of Applications Submitted to DOE to Export LNG.

DOE Reviews Each Application Separately and Considers a Range of Factors to Determine if Approval Is in the Public Interest

According to DOE, when determining whether approval of an application is in the public interest, DOE focuses on (1) the domestic need for natural gas, (2) whether the proposed export threatens the security of domestic natural gas supplies, and (3) whether the arrangement is consistent with DOE's policy of promoting market competition along with other factors bearing on the public interest, such as environmental concerns.[37] In passing the NGA, Congress did not define "public interest;" however, in 1984, DOE developed policy guidelines establishing criteria that the agency uses to evaluate applications for natural gas imports.[38] The guidelines stipulate that, among other things, the market—not the government—should determine the price and other contract terms of imported natural gas. In 1999, DOE began applying these guidelines to natural gas exports.[39]

DOE's review of export applications is not a standardized process, according to agency officials; rather, it is a case-by-case deliberation, where each application is considered separately from others.[40] DOE's review process begins when an applicant submits documentation to DOE requesting permission to export LNG. DOE examines applications one at a time, and it issues a notice of application in the Federal Register to invite persons interested in the application to comment, protest, or intervene.[41] Applicants are then given an opportunity to respond to comments. DOE's internal review includes an examination of the application and analysis of public interest using public comments and applicant responses, the criteria outlined in its policy guidelines, the NGA, DOE's study of the effects of additional LNG exports, and past DOE authorizations. As discussed above, the NGA authorizes DOE to attach terms and conditions necessary to protect the public interest. To further inform its public interest review, DOE commissioned the study of the potential effects of additional exports on the economy.[42] Since the study was released in December 2012, DOE has used it to support its public interest review for each of its application approval documents, including referencing the study's conclusion that LNG exports would have a net positive effect on the economy.[43]

After considering the evidence, DOE issues an order denying the application or granting the application on condition of a satisfactory completion of the NEPA review by FERC.[44] DOE includes the reasoning behind its decision in each order. DOE may also modify the request in an order, such as by limiting the approved export amount or duration.[45] Once DOE conditionally approves an application, it does not grant a final approval until it has reviewed FERC's NEPA document and reconsidered its public interest determination in light of relevant environmental information. Under NEPA, DOE must give appropriate consideration to the environmental effects of its decisions; FERC's NEPA document provides the basis for this consideration.[46]

On May 29, 2014, DOE proposed that it would no longer approve applications for export on the condition of a satisfactory completion of the environmental review by FERC and the resulting NEPA document; instead, it would review applications for export only after FERC issues the NEPA document.[47] On August 15, DOE issued a final decision to implement this procedural change, suspending its practice of issuing conditional orders.[48] DOE also announced on May 29 that it would request another economic study to further examine whether additional exports are in the public interest.[49] At the same time, DOE released two reports on environmental issues for public review—an action prompted by public comments and the number of applications DOE has received since 2010.[50] DOE officials said that, after DOE receives and responds to public comments on these reports, it will use the reports and NEPA document to complete its environmental responsibilities under NEPA.

FERC HAS APPROVED 3 FACILITY APPLICATIONS AND IS REVIEWING 14 OTHERS USING A TECHNICALLY COMPLEX PROCESS THAT INVOLVES OTHER AGENCIES

Since 2010, FERC approved 3 facility applications, including 2 in 2014, and is currently reviewing 14 applications. FERC's reviews of LNG export facility applications are a multiyear analysis of the potential environmental and safety effects of the facility that involves other federal, state, and local agencies.

FERC Approved 3 Facility Applications and Is Reviewing 14 Applications

FERC approved applications to construct and operate the Sabine Pass LNG export facility in April 2012, the Cameron facility in June 2014, and the Freeport facility in July 2014. As of late August 2014, FERC was reviewing 14 applications (See fig. 3). FERC has issued three final NEPA documents in 2014, including for the Cameron and Freeport facilities, and expects to complete one more by the end of 2014.[51] FERC officials said that they could not discuss when the Commission would act on these facility applications.[52]

As shown below, FERC's review of applications to construct LNG export facilities can take 2 to 3 years or more.[53] FERC officials said that these reviews are lengthy because of the complexity of the facilities and number of permits and reviews required by federal and state law. For example, applicants must model the effects of LNG spills from pipes and storage tanks on areas around the facility under a variety of scenarios. One of the applicants we spoke with said that the number of variables involved in modeling a single scenario could require up to a week of computer processing.

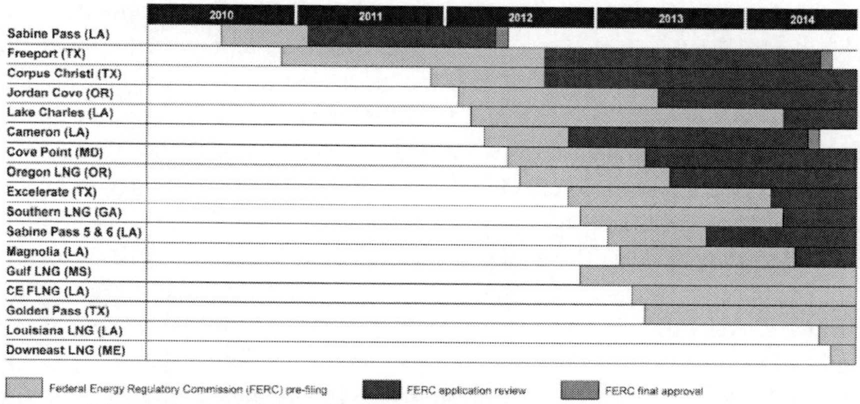

Source: GAO analysis of FERC data. GAO-14-762.

Note: FERC also approved an application to construct and operate a natural gas liquefaction and storage facility in 2008. This facility was identified as the source of LNG to be exported in containers under the Carib Energy application to DOE.

Figure 3. Status of Applications Filed with FERC to Build and Operate LNG Export Facilities.

FERC's Reviews of Applications to Construct LNG Export Facilities Are Technically Complex

FERC's review process is technically complex and includes the following three phases.

Pre-filing. According to FERC officials, the pre-filing phase is intended to allow applicants to communicate freely with FERC staff and stakeholders to identify and resolve issues before the applicant formally files an application with FERC.[54] Under Commission regulations issued pursuant to the Energy Policy Act of 2005, applicants are required to pre-file with FERC a minimum of 6 months before formally filing. The pre-filing phase can vary significantly depending on project specifics; the Freeport and Lake Charles applications were in the pre-filing phase for over 19 months, while the Cameron application was in the pre-filing phase for about 7 months. FERC officials said that the duration of each phase can vary depending on the site specific characteristics of the proposed facility and responsiveness of the applicant to requests for information from FERC.[55] The pre-filing period also involves public outreach by the applicant and FERC, and FERC allows public comments during this period.[56] An applicant completes the pre-filing period when it has submitted the required documentation to FERC and formally filed. This documentation includes a series of 13 resource reports that consist of, among other things, detailed information on project engineering and design, air and water quality, and fish and wildlife, as well as a description of the anticipated environmental effects of the project and proposed mitigation measures. One applicant told us that the resource reports it submitted to FERC consisted of over 12,000 pages.

Application review. The application review phase includes FERC's review of the application and development of the environmental document required by NEPA. FERC officials told us that they start the review phase submits an application.[57] FERC reviews, among other things, facility engineering plans and safety systems identified by the applicant; environmental effects from the construction and operation of the facility; and, potential alternatives to the proposed project.[58] FERC develops a NEPA document with input from relevant agencies that elect to participate, called cooperating agencies, as well as other stakeholders.[59] FERC officials told us that, depending on the location of the proposed facility and amount of construction, FERC prepares either an environmental impact statement (EIS) or environmental assessment (EA). FERC will prepare an EA if it believes the review will find no significant impact on the environment from the project. For example, FERC prepared an EA for the Sabine Pass facility because the proposed facility was within the footprint of an existing LNG import facility and previously the subject of an EIS.[60] FERC officials told us that the agency generally prepares an EIS for proposed facilities that would extend beyond the footprint of an existing import facility. After an EIS or EA is drafted, FERC solicits comments from federal agencies and the public on the document.[61] FERC reviews agency and public comments and integrates those into a final EIS or EA, as necessary. The final EIS or EA will recommend any environmental and safety mitigation measures to be completed during various stages of the project. FERC staff submits the final NEPA document and other staff analyses to FERC commissioners for consideration.[62] FERC commissioners consider the entire record of the proceeding, including the NEPA document, to determine whether to approve a project.

Post-authorization. The post-authorization phase includes FERC oversight of plant construction and operations. After FERC approves a project but before an applicant can start construction, the applicant must develop a plan describing how it will meet any conditions and mitigation measures identified in FERC's approval. FERC oversees construction and ensures that these conditions are met. The Coast Guard and DOT also oversee construction to ensure compliance with their respective regulations. FERC conducts compliance and site inspections during construction at least every 8 weeks. Following construction, the applicant must receive written authorizations from the Commission to begin operations at the facility. Once the facility is operational, FERC conducts annual inspections and requires semiannual status reports from the facility operator.

FERC's Review Process Involves Other Federal, State, and Local Agencies

As the lead agency responsible for the environmental and safety review of LNG export facilities under NEPA, FERC works with federal, state, and local agencies to develop the NEPA document. In some cases, such as with the Corps and DOE, agencies will adopt and use the NEPA document in issuing their respective permits related to the export facility. In addition, FERC regulations require applicants to consult with the appropriate federal, state, and local agencies to ensure that all environmental effects are identified.[63] FERC ensures that the applicant obtains the appropriate federal permits or consultations with these agencies. Major federal participants in FERC's LNG facility review include the following:

- *Coast Guard.* The Coast Guard requires applicants to assess the effects of a new facility on a bordering waterway. The applicant provides the assessment to the Coast Guard for validation and review before filing its FERC application, and the Coast Guard advises FERC on the suitability of the waterway for the LNG marine traffic associated with the facility. The Coast Guard and DOT also assist FERC's review of safety and security of the facility.[64]
- *PHMSA.* PHMSA is an agency within DOT responsible for establishing national policy relating to pipeline safety and hazardous material transportation, including the authority to establish and enforce safety standards for onshore LNG facilities. To assist FERC's assessment of whether a facility would affect public safety, FERC regulations require applicants to show that their facility design would comply with PHMSA regulations for hazardous liquids vapor dispersion and fires.[65] Applicants submit models of vapor dispersion to FERC, and FERC consults with PHMSA to ensure that the models comply with PHMSA regulations.[66]

- *The Corps.* Under section 404 of the Clean Water Act, operations that discharge dredged or fill material into U.S. waters are required to obtain a permit from the Corps.[67] Discharges under this permit must have a state certification to ensure the discharge meets water quality standards.[68] In addition, under section 10 of the Rivers and Harbors Act of 1899, the Corps has regulatory authority to oversee construction activities within the navigable waters of the United States, and applicants may be required to obtain a permit from the Corps.[69]
- *Environmental Protection Agency* (EPA). Applicants may be required under the Clean Air Act (CAA) to receive air permits for the construction and operation of LNG facilities.[70] State environmental agencies generally issue these permits, but EPA can issue the permits if a state is not authorized to issue permits, or under other limited circumstances.[71] EPA also comments on the FERC draft and final EIS, as required by the CAA.[72]

Applicants may also be required by law to consult with these and other federal agencies, such as the National Oceanic and Atmospheric Administration and the Fish and Wildlife Service, to ensure their applications comply with federal laws such as the Endangered Species Act,[73] the Migratory Bird Treaty Act,74 the Magnuson-Stevens Fishery Conservation and Management Act,[75] and the Fish and Wildlife Coordination Act.[76]

In addition to federal permits and consultations, applicants may also be required to obtain other permits under state and local law. Because of the wide variety of projects, locations, and state and local laws, permitting requirements vary by project. The applicant is responsible for identifying the necessary permits and consultations and reporting these to FERC as part of the pre-filing process. In addition to issuing most air permits and water quality certifications, states and local agencies have other permitting and consultation responsibilities, such as to consult with applicants to ensure compliance with the Coastal Zone Management Act[77] and the National Historic Preservation Act.[78]

Agency Comments

We provided a draft of this product to FERC and the DOE for their review and comment. DOE and FERC provided technical comments, which we incorporated throughout the report as appropriate.

Sincerely yours,
Frank Rusco
Director, Natural Resources and Environment

Appendix I: Application Names Used by the GAO

For the purposes of this report, GAO developed table 1 below to allow us to use a single name to refer to related applications to the Federal Energy Regulatory Commission (FERC) and the Department of Energy (DOE). Table 1 lists (1) the names of applicants that submitted requests to FERC to construct liquefied natural gas (LNG) export facilities, (2) the names of applicants that submitted requests to DOE to export LNG from those facilities[1] , and (3) the name GAO used to refer to these applications. In some cases, multiple companies filed jointly for one application.

Table 1. Names of Applications Referred to In This Report, Based on Associated Applications to FERC and DOE

FERC applicant name	DOE applicant name	GAO application name
Sabine Pass[a]	Sabine Pass Liquefaction LLC	Sabine Pass (LA)
Freeport LNG Development LP	Freeport LNG Expansion LP & FLNG Liquefaction LLC	Freeport (TX)
Trunkline LNG Export LLC & Trunkline LNG Company LLC[b]	Lake Charles Exports LLC	Lake Charles (LA)
Dominion Cove Point LNG LP	Dominion Cove Point LNG LP	Cove Point (MD)
N/A[c]	Freeport LNG Expansion LP & FLNG Liquefaction LLC	Freeport Expansion (TX)
Cameron LNG LLC[d]	Cameron LNG LLC	Cameron (LA)
Jordan Cove Energy Project LP	Jordan Cove Energy Project LP	Jordan Cove (OR)
Corpus Christi Liquefaction LLC	Cheniere Marketing LLC	Corpus Christi (TX)
Oregon LNG[e]	LNG Development Company LLC	Oregon LNG (OR)

Table 1. (Continued)

FERC applicant name	DOE applicant name	GAO application name
Excelerate Liquefactions Solutions	Excelerate Liquefactions Solutions I, LLC	Excelerate (TX)
Elba Liquefaction Company LLC & Southern LNG Company LLC	Southern LNG Company LLC	Southern LNG (GA)
Sabine Pass	Sabine Pass Liquefaction LLC	Sabine Pass Trains 5 & 6 (LA)
Magnolia LNG LLC	Magnolia LNG LLC	Magnolia LNG (LA)
Gulf LNG Liquefaction Company LLC	Gulf LNG Liquefaction Company LLC	Gulf LNG (MS)
CE FLNG LLC & CE Pipeline LLC	CE FLNG LLC	CE FLNG (LA)
Golden Pass Products LLC & Golden Pass Pipeline LLC	Golden Pass Products LLC	Golden Pass (TX)
Louisiana LNG Energy LLC	Louisiana LNG Energy LLC	Louisiana LNG (LA)
Downeast LNG, Inc.	N/A[f]	Downeast LNG (ME)
N/A[g]	Carib Energy (USA) LLC	Carib Energy (FL)

Sources: GAO analysis of FERC and DOE documents. | GAO-14-762

[a] The application was submitted by Sabine Pass Liquefaction Expansion LLC, Sabine Pass Liquefaction LLC, and Sabine Pass LNG LP, to which FERC collectively refers as “Sabine Pass.”

[b] FERC also included in the application the Pipeline Modification Project submitted by Trunkline Gas Company LLC. The project coincides with the construction of the LNG export facility.

[c] There was effectively a single application to FERC to construct the Freeport LNG export facility. DOE received two applications to export LNG for this facility.

[d] FERC also included in the application the Pipeline Project submitted by Cameron Interstate Pipeline LLC. The project coincides with the construction of the LNG export facility.

[e] FERC also included in the application to modify the pipeline route submitted by Oregon Pipeline Company LLC. The project coincides with the construction of the LNG export facility.

[f] As of late August 2014, Downeast LNG had not submitted an application to DOE to export LNG.

[g] The Carib Energy application to DOE requested approval to export LNG via specialized container. The LNG will be processed and loaded into containers at a small liquefaction and storage facility that has multiple uses, and is not identified by FERC as an LNG export facility.

End Notes

[1] Section 3 of the Natural Gas Act (NGA) (15 U.S.C. § 717b) requires federal approval of natural gas exports. For crude oil exports, the Energy Policy and Conservation Act of 1975 (42 U.S.C. § 6212(b)) directs the President to "promulgate a rule prohibiting the export of crude oil" produced in the United States. The Department of Commerce administers export licenses for crude oil.

[2] The Kenai LNG export facility in Nikiski, Alaska, has been authorized to export LNG to Japan for the last 45 years. According to DOE, because there is no pipeline interconnection between Alaska and the lower 48 states, those LNG markets are generally viewed as distinct.

[3] Pub. L. No. 75-688, §3, 52 Stat. 822, codified as amended at 15 U.S.C. § 717b.

[4] In 1992, Congress amended the NGA to require DOE to treat applications to export LNG to FTA countries as consistent with the public interest. Unless otherwise indicated, the DOE applications discussed in this report are non-FTA applications.

[5] In 1984 DOE delegated to FERC the authority to approve or deny LNG facilities. DOE Delegation Order No. 0204-112, 49 Fed. Reg. 6684, 6690 (Feb. 22, 1984). FERC also authorizes the construction and operation of associated pipelines under section 7 of the NGA.

[6] For facilities located offshore beyond state waters, the Maritime Administration (MARAD) within the Department of Transportation authorizes construction pursuant to the Deepwater Port Act. According to a DOE official, MARAD serves as the lead agency for the environmental review of offshore LNG facilities beyond state waters. According to this official, DOE has received two applications to export LNG from these offshore facilities, but MARAD has not received applications to construct the associated facilities.

[7] 79 Fed. Reg. 32261 (June 4, 2014).

[8] 79 Fed. Reg. 48132 (Aug. 15, 2014). Applications that DOE has conditionally approved still require a final approval from DOE after DOE has reviewed FERC's environmental document. The revised process applies only to the lower 48 states.

[9] For the purposes of this report, we will refer to companies that apply to DOE or FERC for export or facility approvals as "applicants." As of late August 2014, in addition to 35 applications to non-FTA countries, DOE has received 42 applications to export LNG to FTA countries. These totals are for the lower 48 states. In July 2014, DOE also received an application to export LNG from a proposed facility in Alaska to FTA and non-FTA countries.

[10] Because these were nonprobability samples, the information collected during site visits cannot be generalized to all LNG facilities but provides illustrative examples.

[11] CRS, U.S. Natural Gas Exports: New Opportunities, Uncertain Outcomes R42074 (April 8, 2013).

[12] There are currently six onshore import facilities on the Gulf Coast and three on the East Coast. In addition, there are two offshore import facilities on the East Coast.

[13] In particular, the application of horizontal drilling techniques and hydraulic fracturing—a process that injects a combination of water, sand, and chemical additives under high pressure to create and maintain fractures in underground rock formations that allow oil and natural gas to flow—have increased U.S. natural gas and crude oil production.

[14] EIA is a statistical agency within the DOE that collects, analyzes, and disseminates independent information on energy issues.

[15] Withdrawals from other natural gas sources such as gas wells, oil wells, and coalbed wells decreased over this period.

[16] From 2007 to 2013, the average annual Henry Hub price of natural gas decreased from $6.97 per million British thermal unit (MMBtu) to $3.73 per MMBtu. The Henry Hub is a major gas trading center in the Gulf of Mexico and the delivery point for natural gas futures contracts on the New York Mercantile Exchange.

[17] In July 2014, FERC estimated the following landed prices of natural gas for August 2014: at the Cove Point import facility in Maryland - $3.27 per MMBtu; in the United Kingdom - $6.59 per MMBtu; in India- $11.20 per MMBtu, and in Japan $11.35 per MMBtu. The landed price represents the price of LNG at an import facility, before it has been converted back to a gas.

[18] LNG can also be stored in specialized containers and transported on cargo vessels. LNG exports by container are generally much smaller volumes than exports on an LNG tanker. For example, Carib Energy, 1 of the 3 export applications approved by DOE, was approved to export 0.04 billion cubic feet per day of natural gas in liquid form in these containers. For comparison, Sabine Pass, the first LNG export application approved by DOE for export by LNG tanker, was approved to export 2.2 billion cubic feet per day.

[19] Natural gas transported by pipeline contains, among other contaminants, small amounts of water, carbon dioxide, hydrogen sulfide, and mercury, as well as hydrocarbons such as pentane and hexane that could freeze and contaminate the liquefaction process.

[20]The United States exports a small amount of LNG that has been shipped to the United States, stored in LNG tanks at import facilities, and then reexported when it makes economic sense for the companies involved.

[21] 15 U.S.C. § 717b.

[22] DOE Delegation Order 0204-112, 49 Fed. Reg. 6684, 6690 (Feb. 22,1984).

[23] Countries that have FTAs with the United States and can trade natural gas without restrictions are Australia, Bahrain, Canada, Chile, Colombia, the Dominican Republic, El Salvador, Guatemala, Honduras, Jordan, Mexico, Morocco, Nicaragua, Oman, Panama, Peru, South Korea, and Singapore.

[24] 15 U.S.C. § 717b(c).

[25] As of late August 2014, DOE has approved 37 applications to export LNG to FTA countries, amounting to nearly 38 billion cubic feet per day of natural gas in the form of LNG. In addition, DOE has approved 10.56 billion cubic feet per day in conditional or final export approvals to non-FTA countries.

[26] Enacted in 1970, NEPA has as its purpose, among others, to promote efforts to prevent or eliminate damage to the environment. NEPA requires an agency to prepare a detailed statement on the environmental effects of any "major federal action" significantly affecting the environment. Regulations promulgated by the Council on Environmental Quality implementing NEPA generally require an agency to prepare either an environmental assessment (EA) or an environmental impact statement (EIS), depending on whether or not a proposed action could significantly affect the environment.

[27] 15 U.S.C. § 717b-1(a).

[28] DOE's review is also considered a federal action and subject to NEPA, and DOE uses FERC's NEPA document and its own analysis to fulfill its NEPA responsibility.

[29] FERC is overseen by a commission of up to five Presidential appointees who serve 5- year terms.

[30] DOE also notes that U.S. LNG exports will diversify international supply options and improve energy security for many allies and trading partners of the United States.

[31] Opponents have identified numerous concerns related to hydraulic fracturing, including the potential for groundwater contamination.

[32] The Sabine Pass application was fully approved by DOE in August 2012 following FERC's facility approval in April 2012. FERC approved applications to construct the Cameron and Freeport facilities in June and July of 2014, respectively. According to FERC officials, the Cameron facility was given notice to proceed with site preparation activities for its facility in July 2014. LNG exports under the Carib Energy application will not pass through an LNG export facility; instead, LNG will be processed at a natural gas liquefaction and storage facility where it will be loaded into specialized containers and transported by truck to various ports for export on cargo vessels.

[33] When DOE conditionally approved the Sabine Pass application in May 2011, it had received 2 other LNG export applications and expected more. DOE received 13 other nonFTA applications during the period of the two-part study.

[34] The study consisted of two parts, a study from the EIA and a study from NERA Economic Consulting (NERA). EIA study, Effect of Increased Natural Gas Exports on Domestic Energy Markets; NERA study, Macroeconomic Impacts of LNG Exports from the United States.

[35] DOE conditionally approved applications to export from the following facilities: Freeport, Texas (two separate applications); Lake Charles and Cameron, Louisiana; Cove Point, Maryland; and Jordan Cove and Oregon LNG, Oregon.

[36] DOE issued a final approval for the Carib Energy application because it found that FERC had already completed a NEPA review of the LNG liquefaction and storage facility that Carib Energy proposed to use; DOE determined that FERC's environmental review addressed all of the reasonably foreseeable activities likely to arise from Carib Energy's proposed use of the facility. More specifically, DOE determined that the Carib Energy application qualified for a categorical exclusion, which it issued on May 30, 2014. An agency may generally exclude categories of actions from the requirement to prepare a NEPA document if it has determined that the actions do not individually or cumulatively have a significant impact on the environment and that there are no extraordinary circumstances that might cause significant environmental effects.

[37] See Sabine Pass Liquefaction LLC, FE Docket No. 10-111-LNG, DOE/FE Order No. 2961, p. 27-29; May 20, 2011.

[38] 49 Fed. Reg. 6684 (Feb. 22, 1984).

[39] Phillips Alaska Natural Gas Corporation and Marathon Oil Company. DOE Order No. 1473, Order Extending Authorization to Export Liquefied Natural Gas from Alaska (Apr. 2, 1999).

[40] DOE officials said that, although reviews are handled on a case-by-case basis, each deliberation considers the cumulative effects from prior approvals.

[41] Any member from the public can comment on or protest applications during a 60-day comment period. In addition, an entity can file a motion or notice to intervene with DOE. Intervenors become participants in the proceeding and can request a rehearing after DOE issues an order. DOE officials examined and considered almost 200,000 public comments on its commissioned study in its review of the applications for Freeport, Lake Charles, Cameron, Cove Point, Jordan Cove, and Oregon LNG. When reviewing public comments, DOE officials said they consider the merits of the issue addressed, not the volume of comments submitted.

[42] 77 Fed. Reg. 73627 (Dec. 11, 2012). The study consisted of two parts. The first part of the study was conducted by EIA and examined the potential effect of 6 to 12 billion cubic feet

per day of natural gas exports on domestic energy consumption, production, and prices under various scenarios. This part of the study found that natural gas prices would rise gradually under most study conditions. The second part of the study, conducted by NERA, evaluated the macroeconomic effects of LNG exports on the U.S. economy using the results of the EIA study. This part of the study found that LNG exports would have net economic benefits in all cases. One concern of export opponents was that NERA used data from EIA's 2011 Annual Energy Outlook, which did not reflect more recent EIA reports that projected greater increases in domestic demand for natural gas. In February 2014, in response to a request from Cheniere, NERA updated its study using data from EIA's 2013 Annual Energy Outlook and found that its conclusions from the original study were still valid.

[43] Additionally, in December 2012, DOE established a precedence order for reviewing applications. Under this order, DOE prioritized applications from companies that had already entered the filing process with FERC, and moved those applications ahead of others that had not started the process with FERC.

[44] According to a DOE official, MARAD would lead the NEPA review for offshore facilities beyond state waters.

[45] DOE conditionally approved an export application for Freeport for 0.4 billion cubic feet per day of natural gas rather than the 1.4 billion cubic feet per day that the applicant requested. DOE approved the lesser quantity because the facility application Freeport submitted to FERC was not capable of handling the full amount of natural gas proposed to DOE. that, after DOE receives and responds to public

[46] DOE is a cooperating agency for FERC's NEPA review of LNG export projects and contributes to the development of the NEPA document.

[47] 79 Fed. Reg. 32261 (June 4, 2014). The change would also supersede the precedence order. According to the DOE notice, DOE could still choose to implement the policy of issuing conditional orders at a later date. According to DOE officials, this change would allow them to use agency resources more efficiently because they would conduct a single review of each application instead of separate reviews for conditional and final approvals. In addition, the proposal would allow projects that are more commercially advanced to be reviewed by DOE once FERC has issued a NEPA document.

[48] 79 Fed. Reg. 48132 (Aug. 15, 2014). According to DOE officials, the benefits of this change include DOE being able to ensure prompt action on applications that are otherwise ready to proceed, and to better allocate departmental resources by reducing the likelihood that the DOE would be forced to act on applications with little prospect of proceeding.

[49] According to the DOE notice, DOE requested that the EIA update its 2012 study, which had been included as part 1 of DOE's LNG export study. DOE requested that EIA evaluate the effect of increased natural gas demand—reflecting possible exports of U.S. natural gas—on domestic energy markets using the modeling analysis presented in its 2014 Annual Energy Outlook as a starting point. DOE officials said that they commissioned the new study to include updated data and consider export scenarios ranging from 12 to 20 billion cubic feet per day. In addition to the EIA study, DOE officials said they planned to commission an outside contractor to conduct a macroeconomic review using EIA's results.

[50] 79 Fed. Reg. 32258, 32260 (June 4, 2014). The goal of the two studies, according to the DOE notice, was to provide the public with further information on the 1) effect of LNG exports on greenhouse gas emissions and 2) potential environmental impacts from increased development of unconventional natural gas resources, particularly production that involves hydraulic fracturing.

[51] FERC has issued three final NEPA documents in 2014. These include the Cameron final EIS, issued April 30, 2014 (facility approved by the Commission on June 19, 2014); the Freeport final EIS, issued June 16, 2014 (facility approved by the Commission on July 30, 2014); and the Cove Point EA, issued on May 15, 2014. FERC also issued a draft EIS for Corpus Christi on June 13, 2014.

[52] FERC staff are restricted by regulation from discussing the nature or timing of future Commission actions. 18 C.F.R. §3c.2(b) 2014.

[53] Application estimates based on the duration of FERC's completed and pending applications.

[54] Once an applicant formally files, FERC ex parte regulations require all communications between an applicant and FERC to be on the record.

[55] FERC officials told us that the applicant generally determines how long pre-filing will last. After the 6-month minimum pre-filing period has been reached and the applicant has submitted the appropriate documentation, FERC staff cannot prevent the applicant from ending the pre-filing phase by filing an application.

[56] According to FERC officials, a public scoping period typically occurs for 30 days after FERC issues a notice in the Federal Register announcing its intent to prepare an environmental document. The goal of this period is to determine the scope of issues to be addressed and to identify significant issues related to a proposed action.

[57] According to FERC officials, the public has the opportunity to intervene and become a party to the proceeding for a period specified in FERC's notice in the Federal Register announcing that an application has been filed (typically 21 days). Parties have the right to seek court review of Commission orders.

[58] For example, one aspect of this review includes an analysis of liquefaction systems, fire suppression and alarm systems, instrumentation, transfer piping, and dikes and sumps to ensure that the effects of any LNG spill stay within the facility boundary.

[59] Cooperating agencies have jurisdiction by law or special expertise with respect to resources potentially affected by the proposal and participate in the NEPA analysis. According to FERC officials, the Coast Guard, the Corps, DOT, DOE, and the Environmental Protection Agency are typically cooperating agencies for FERC LNG facility reviews. According to these officials, a state or local agency or American Indian tribe may also become a cooperating agency under certain conditions.

[60] FERC must conduct an EIS after an EA if the EA determines that significant impacts are likely.

[61] FERC issues a draft and final document for an EIS. According to FERC officials, the public has the opportunity to comment, typically for 45 days after FERC issues a draft EIS, and FERC addresses those comments in the final EIS. EA's do not include a draft document; the public has the opportunity to comment, typically for 30 days after the EA is completed. FERC addresses these comments in the Commission order. The public may also file a motion to intervene during these comment periods.

[62] After a final EIS or EA is issued, federal agencies or agencies with delegated federal responsibilities are to issue a final decision on the applicant's request for the federal authorization within 90 days, unless a schedule is otherwise established by federal law. 18 C.F.R. §157.22.

[63] 18 C.F.R. §380.3 (b)(3),(4).

[64] In 2004, to help ensure coordination between federal agencies, FERC, DOT, and the Coast Guard entered an interagency agreement that defined responsibilities among the agencies regarding the safety and security at waterfront LNG facilities.

[65] Some applicants told us that, in August 2013, PHMSA asked them for more information for its review of vapor models than previously required, and that PHMSA took more time to review the models. Some applicants also told us that staffing shortages at PHMSA caused delays in these reviews. PHMSA officials we spoke to said staff has been added to address these delays.

[66] PHMSA has indicated that these regulations are out of date and do not reflect modern modeling technology. PHMSA officials told us that the agency plans to revise these regulations accordingly, and that it has issued guidance to help with interpretation of these regulations.

[67] 33 U.S.C. §1344. EPA may prohibit, withdraw, deny, or restrict section 404 permits.

[68] 33 U.S.C. §1341. EPA may review the water quality certifications issued by state agencies.

[69] 33 U.S.C §403. The Rivers and Harbors Act regulates any work or structures that potentially affect the navigable capacity of a body of water.

[70] According to EPA, potentially applicable CAA requirements include prevention of significant deterioration, nonattainment new source review (NSR), minor source NSR, title V, new source performance standards, hazardous air pollutants, and other state air regulations.

[71] The majority of states are authorized or approved to implement the relevant CAA permitting programs relevant to LNG facilities.

[72] 42 U.S.C. §7609(a).

[73] The Endangered Species Act, Pub. L. No. 93-205, codified as amended at 16 U.S.C. §§1531-1543, has as its purpose, to conserve threatened and endangered species and the ecosystems upon which they depend. Under section 7 of the act, federal agencies must ensure that any action they authorize, fund, or carry out is not likely to jeopardize the continued existence of a species protected under the act.

[74] The Migratory Bird Treaty Act, act of July 3, 1918, ch. 128, codified as amended at 16 U.S.C. §§ 703-712, implements various treaties and conventions between the United States, Canada, Japan, Mexico, and the former Soviet Union for the protection of migratory birds.

[75] The Magnuson-Stevens Fishery Conservation and Management Act, Pub. L. No. 94– 265 (1976), codified as amended at 16 U.S.C. §1801 et seq., provides the statutory framework for the protection and management of the nation's marine fishery resources.

[76] The Fish and Wildlife Coordination Act, Act of March 10, 1934, ch. 55, codified as amended at 16 U.S.C. §§ 661-666 provides the basic authority for the Fish and Wildlife Service's involvement in evaluating impacts to fish and wildlife from proposed water resource development projects.

[77] The Coastal Zone Management Act of 1972, Pub. L. No. 92-583 (1972), codified as amended at 16 U.S.C. §§ 1451-1466, promotes comprehensive and coordinated planning for coastal zone development and preservation between states and the federal government.

[78] The National Historic Preservation Act, Pub. L. No. 89-665 (1966), codified as amended at 16 U.S.C. §§ 470 to 470x-6 requires, among other things, that federal agencies consider the effects of any federally funded or permitted project on any historic site, building, structure, or other object that is listed on the National Register of Historic Places.

End Note for Appendix I

[1] DOE has received over 20 other applications to export LNG from companies that have not filed with FERC and are not listed here.

In: U.S. Exports of Oil and Natural Gas
Editor: Frank G. Adams
ISBN: 978-1-63463-904-0

Chapter 3

U.S. EXPORTS OF CRUDE OIL AND NATURAL GAS: THE CASE OF ALASKA*

Paul W. Parfomak and Ian F. Fergusson

SUMMARY

Recent growth in U.S. natural gas and crude oil production has fundamentally shifted the energy supply and demand balance in key U.S. energy markets. Many analysts and energy producers argue that there is currently an oversupply of natural gas in U.S. gas-producing regions, and that supplies of certain light sweet Western crudes (notably from the Bakken region) exceed the demand of Gulf Coast refineries. The result has been a spate of applications by the U.S. natural gas industry for federal permits to export natural gas to overseas buyers. Such exports are permitted by statute, subject to a national interest determination by the Department of Energy. Perceived oversupply of Bakken crude oil has similarly led to calls by some market stakeholders and policy makers for a relaxation of the federal restrictions on crude oil exports. Numerous legislative proposals in the 113th Congress, such as the Natural Gas Export Promotion Act of 2014 (S. 2494) and the Crude Oil Export Act (H.R. 4349), would facilitate increased exports of domestic energy resources.

* This is an edited, reformatted and augmented version of a Congressional Research Service publication R43753, prepared for Members and Committees of Congress, dated October 1, 2014.

Whether—or to what extent—the United States should allow overseas exports of crude oil and natural gas is the subject of ongoing Congressional debate. However, U.S. exports of both natural gas and crude oil are not new. Congress and prior presidential administrations have authorized such exports before in circumstances where they were viewed as beneficial to the United States. The case of Alaska is of particular relevance because it is a major energy producer and has a history of both crude oil and natural gas exports going back several decades. Exports of crude oil from Alaska are currently authorized by three different statutes pertaining to (1) crude transported via the Trans Alaska Pipeline System, (2) Cook Inlet crude, and (3) crude oil exported to Canada under the North American Free Trade Agreement. Export of Cook Inlet natural gas from the Kenai LNG terminal has been authorized repeatedly since 1969. Export of natural gas from Alaska's North Slope was authorized in 1988, although the applicability of this authorization to the proposed Alaska LNG Project has not been resolved.

In the context of the current export debate, a review of Alaskan crude oil and natural gas exports may offer historical perspectives of value to policy makers. Alaska's experience shows that— under a specific set of circumstances—such exports have been viewed as in the national interest both by Congress and successive presidential administrations. Regional production trends, market prices, and local commodity demand have all been important considerations in establishing export policies for Alaska. It is also instructive to note that, even with long-standing export approvals in effect, exports of crude oil and natural gas from Alaska have been relatively modest. Production economics and competitive market forces will be the ultimate determinant of export volumes where exports are permitted.

In the context of broader U.S. oil and natural gas export policy the Alaska experience may raise several key questions. To what extent does the rationale for energy exports from Alaska—which is geographically isolated—apply to other U.S. supply regions? What is the interplay between overseas exports and the maritime shipping industry? What are the expectations for capital investment by developers to support export production and how might they affect the nation's overall oil and gas supplies? What are the environmental impacts of increased production for export? How might international trade agreements influence the U.S. government's ability to tailor its oil and natural gas export policies by region? As Congress continues its oversight of the nation's energy resources, understanding these issues may be important.

INTRODUCTION

The deployment of shale energy technology, including hydraulic fracturing and horizontal drilling, in the United States has created a new abundance of crude oil and natural gas in a number of energy-producing regions across the country. Since 2008, U.S. production of crude oil from the lower-48 states has more than doubled, from 3.1 million barrels per day (bbl/d) to 6.6 million bbl/d in 2014.[1] Likewise, over the last ten years, lower-48 states' natural gas production has risen from 41.1 billion cubic feet per day (bcf/d) to 69.8 bcfd.[2] These production trends are expected to continue.

The growth in domestic natural gas and crude oil production has fundamentally shifted the energy supply and demand balance in key U.S. energy markets. Many analysts and energy producers argue that there is currently an oversupply of natural gas in U.S. gas-producing regions (with some important exceptions), and that supplies of certain light sweet Western crudes (notably from the Bakken region of Montana and North Dakota) exceed the demand of Gulf Coast refineries, which are configured primarily to process heavy crudes. The result has been a spate of applications by the U.S. natural gas industry for federal permits to export natural gas to overseas buyers.[3] Such exports are permitted by statute, subject to a national interest determination by the Department of Energy.[4] Perceived oversupply of Bakken crude oil has similarly led to calls by some market stakeholders and policy makers for a relaxation of the federal restrictions on crude oil exports.[5] Numerous legislative proposals in the 113th Congress, such as the Natural Gas Export Promotion Act of 2014 (S. 2494) and the Crude Oil Export Act (H.R. 4349), would facilitate increased exports of domestic energy resources.

Whether—or to what extent—the United States should allow overseas exports of crude oil and natural gas is the subject of ongoing Congressional debate. However, it may be noted that U.S. exports of both natural gas and crude oil are not new. Congress and prior presidential administrations have authorized such exports before in circumstances where they were viewed as beneficial to the United States. The case of Alaska is of particular relevance because it is a major energy producer and has a history of both crude oil and natural gas exports going back several decades. In the context of the current export debate, a review of Alaska's experiences with crude oil and natural gas exports may offer historical perspectives of value to policy makers.

This report provides a summary overview of Alaskan crude oil and natural gas export authorizations since the late 1960s, including a review of export activity. The report does not address exports of refined products (e.g., diesel), natural gas liquids, or coal.[6]

Energy Development in Alaska

Alaska's principal energy-producing basins are located on the North Slope in the Arctic and in the Cook Inlet region to the south (**Figure 1**). Development of North Slope oil began with the discovery of the Prudhoe Bay oil field in 1968 and the subsequent development of the Trans Alaska Pipeline System (TAPS) in the 1970s to deliver North Slope crude, as well as smaller quantities of natural gas liquids (NGLs), such as propane, to a marine terminal at Valdez on the state's southern coast. As discussed below, the North Slope contains significant dry natural gas resources as well, but these have not been commercially developed beyond serving local needs.

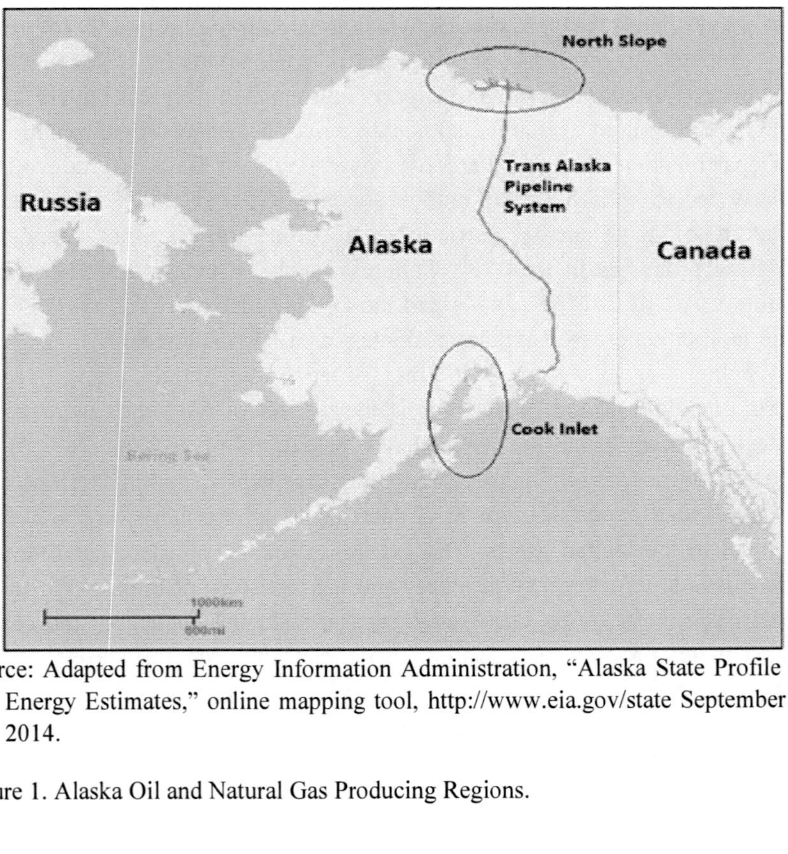

Source: Adapted from Energy Information Administration, "Alaska State Profile and Energy Estimates," online mapping tool, http://www.eia.gov/state September 26, 2014.

Figure 1. Alaska Oil and Natural Gas Producing Regions.

Crude oil drilling in the Cook Inlet region has a much longer history, beginning in the early 1900s and continuing (sporadically) through World War II. However, large scale commercial development did not begin until the late 1950s after discoveries by Richfield Corporation at Swanson River.[7] Natural gas development in Cook Inlet began in 1959 with three gas wells drilled by Union Oil Company of California and Ohio Oil Company near Kenai.[8] As does the North Slope, Cook Inlet produces commercial volumes of NGLs associated with both its oil and natural gas operations.

Oil Exports from TAPS

The export of crude oil from Alaska's North Slope (ANS) has been an issue for Congress since the authorization of the Trans Alaska Pipeline System in 1973. The Trans Alaska Pipeline Authorization Act[9] specified that oil shipped through the pipeline could be exported internationally, but only under restrictive conditions (detailed in *Appendix A*) including a finding by the President that such exports would be in the national interest. These conditions were reaffirmed in the Export Administration Act of 1979[10] (*Appendix A*). In effect, the conditions amounted to a ban on TAPS oil exports.

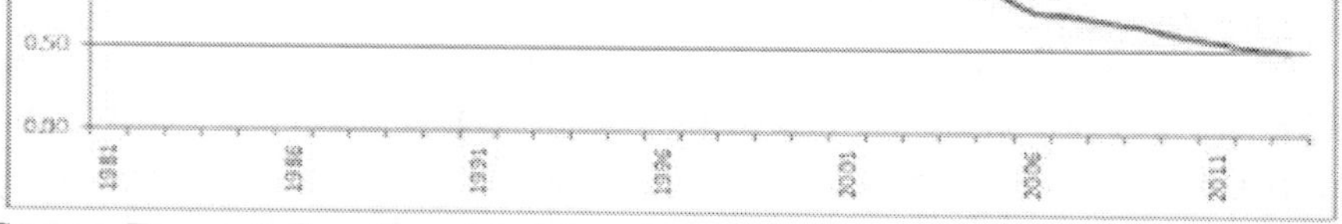

Source: Energy Information Administration (EIA), "Alaska North Slope Crude Oil Production," web page, September 29, 2014, http://www.eia.gov/dnav/pet /hist/ LeafHandler.ashx?n=PET&s=MANFPAK1&f=M.

Notes: EIA does not report production prior to 1981.

Figure 2.Alaska North Slope Crude Oil Production. (Million barrels per day (bbl/d))

TAPS was completed in 1977; initial oil shipments were flowing by year-end. With continued oilfield development on the North Slope, production climbed steadily for 10 years, peaking at nearly 2 million barrels per day (bbl/d) in 1988 (*Figure 2*). Much of the North Slope crude was shipped to California for refining, which was the nation's third-largest oil producer at the time.

The United States-Canada Free-Trade Agreement Implementation Act of 1988 (USCFTA)[11] authorized exports of up to 50,000 bbl/d[12] of TAPS crude oil to Canada—as provided for in the agreement,[13]—subject to certain presidential findings and on condition that the oil be shipped on U.S.-flagged tankers from the lower 48 states and consumed in Canada.[14] It appears that access to TAPS oil was a Canadian request, based on the needs of Vancouver refineries. Lacking access to TAPS crude, Canadian refiners on the west coast had to source their oil feedstock from more expensive Asian or Middle Eastern suppliers. Canada sought this reciprocal access for its acquiescence to essentially guarantee a supply of oil and gas to the United States under the USCFTA energy chapter. It was noted that 50,000 bbl/d was only 2-3% of total Alaskan daily output in 1988.[15] To implement this provision of the agreement, on December 31, 1988, President Reagan issued a finding that crude oil exports under USCFTA were in the national interest.[16] According to the Alaska Department of Natural Resources (ADNR), however, there were no shipments of TAPS crude to Canada under this authorization.[17]

In the mid-1990s, high volumes of Alaskan oil could be shipped economically only to the four western states.[18] This resulted in locally falling oil prices and constrained domestic production.[19] As California prices fell below the world market, there were complaints from West Coast oil producers, industry analysts, and public officials about what they perceived to be artificially depressed prices. Some also pointed to the underutilization of tankers built to carry North Slope crude, and the resulting negative impacts on marine employment, shipbuilding and repair, and the availability of tankers for national defense.[20]

Early efforts to achieve remedial action failed until 1995, when low world oil prices, a relatively modest level of net oil imports (8.0 million bbl/d), and a supportive Department of Energy (DOE) coincided with renewed legislative efforts in both Houses of Congress. A June 1994 DOE study, *Exporting Alaskan North Slope Crude Oil—Benefits and Costs*, concluded that permitting the export of Alaska crude would be beneficial to the U.S. economy.

> First, lifting the ban would expand the markets in which ANS oil can be sold, thereby increasing its value. ANS oil producers, the States of California and Alaska, and some of their local governments all would benefit from increased revenues. Permitting exports also

> would generate new economic activity and employment in California and Alaska. The study concludes that these economic benefits would be achieved without increasing gasoline prices (either in California or in the nation as a whole).[21]

Increased producer revenues would be the result of access to a broader market as well as transportation savings realized by avoiding a trip through the Panama Canal to reach U.S. ports in the Gulf of Mexico. In addition, DOE predicted that higher prices at the wellhead would result in 100,000 bbl/d more output from Alaska and California than would be the case with continued export restriction. Higher North Slope production, in turn, would generate additional income to the federal government due to oil sales from federally owned reserves and royalties on federal leases. Royalty revenues to the state of Alaska would likewise increase.

Original opposition to the export of crude oil from TAPS was driven in part by representatives of the U.S. maritime industry, who viewed Alaskan oil development as an opportunity to enlarge the U.S.-flagged coastal tanker fleet under the Merchant Marine Act of 1920 (the Jones Act).[22] The Jones Act requires shipments between U.S. ports to be carried on U.S.-flagged vessels. More recent opponents of exports from TAPS argued that such exports would increase U.S. dependence upon foreign oil supplies, raise gasoline prices, and lead to job losses at West Coast refineries no longer being supplied with Alaskan oil. Some opponents also expressed concerns about potential oil spills (in light of the *Exxon Valdez* spill) or linked TAPS oil exports to potential oil exploration in Alaska's Arctic National Wildlife Refuge, an environmentally sensitive area they wished to protect from development.[23]

Notwithstanding the arguments of opponents, and with substantial projected benefits (and little administrative cost), export ban repeal bills in the 104th Congress (H.R. 70 and S. 395) passed by large margins, 324-77 and 74-25 respectively.[24] President Clinton signed the Trans-Alaska Pipeline Amendment Act (P.L. 104-58) in November 1995. The act provides that oil transported through TAPS may be exported unless the President finds, after considering specified criteria, that exports are not in the national interest (*Appendix A*).[25]

To address the economic and environmental issues associated with TAPS oil exports, the National Economic Council, the Council on Environmental Quality, and the Department of Commerce's Bureau of Export Administration coordinated an interagency review. The review—which included extensive public hearings, review of public comments, and analytical evaluation—

concluded that TAPS exports would not likely pose a significant impact to the economy or the environment.[26] Subsequently, on April 28, 1996, President Clinton issued a national interest determination authorizing North Slope oil exports. The President's determination stated that such exports

> will not diminish the total quantity or quality of petroleum available to the United States; and are not likely to cause sustained material oil supply shortages or sustained oil price increases significantly above world market levels that would cause sustained material adverse employment effects in the United States or that would cause substantial harm to consumers.... I have also ... concluded that exports of such crude oil will not pose significant risks to the environment if certain terms and conditions are met.[27]

The oil export authorization included several conditions related to tanker shipping routes, inspections, and ballast exchange intended to mitigate environmental risk.

With the crude oil export restrictions lifted, TAPS exports totaling 36,000 bbl/d began in 1996; they grew to 66,500 bbl/d in 1997, dipped to 52,900 bbl/d in 1998, and rose again to a high of 74,000 bbl/d in 1999—about 7% of North Slope production that year. According to the Energy Information Administration (EIA), Alaskan crude oil exports between 1996 and 2004 were shipped to South Korea (48%), Japan (26%), China (17%), and Taiwan (9%).[28] TAPS exports ceased temporarily in May 2000 as West Coast buyers had to pay more to compete with foreign buyers for Alaskan oil. An additional cargo was shipped to China in 2004 aboard a tanker reportedly en route to Asia for repairs.[29] In September 2014, TAPS exports resumed with the shipment of 800,000 barrels of crude oil aboard a tanker reportedly bound for a refinery in South Korea.[30]

Viewed relative to total domestic refinery input of 14.8 million bbl/d,[31] TAPS crude oil exports at their peak in 1999 amounted to the equivalent of half of one percent of U.S. refinery demand. In absolute terms, these export volumes were not viewed by market analysts as particularly significant. In a July 1999 report, the General Accounting Office (GAO)[32] concluded that lifting the North Slope oil export ban raised the relative prices of North Slope and comparable California oils between $0.98 and $1.30 per barrel above what they would have been with the ban in place, but that the price increases did not have an observable effect on North Slope or California oil production. (It is possible that production could have been lower without the incremental demand for exports.) The GAO also concluded that lifting the export ban increased costs for some refiners but had limited effects on consumers and the oil-shipping industry on the West Coast.[33]

OIL EXPORTS FROM COOK INLET

Exports of Alaskan crude oil other than those passing through TAPS are subject to restrictions under the Energy Policy and Conservation Act of 1975 (P.L. 94-163, EPCA). This act required the President to promulgate a rule prohibiting the export of domestic crude oil, generally, but allowed exemptions for exports which the President determines to be consistent with the national interest and the purposes of the act (§103(b)) as further detailed in *Appendix A*. Responsibility for implementing EPCA has been delegated by the President to the Secretary of Commerce. Therefore, crude oil export licenses under EPCA are issued by the Bureau of Industry and Security (BIS) within the Department of Commerce.

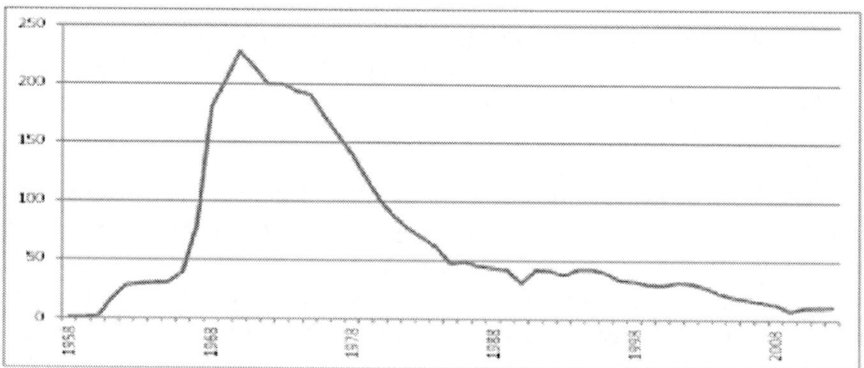

Source: Alaska Department of Natural Resources, Division of Oil and Gas Annual Report 2013, 2013, p. 27.

Figure 3. Cook Inlet Crude Oil and NGL Production. (Thousand barrels per day (bbl/d)).

On November 6, 1985, the Secretary of Commerce determined that the export of crude oil from Alaska's Cook Inlet is consistent with the national interest and the purposes of EPCA. The exemption was granted in part due to the efforts of Alaska's congressional delegation and others to persuade the Reagan Administration that increased oil revenues would spur greater oil development in Alaska in the face of declining production.[34] Crude oil production from Cook Inlet state waters at that time was around 40,000 bbl/d, but was declining approximately 15% per year due to a lack of new investment by incumbent oil producers (*Figure 3*).[35]

In his national interest determination, the Secretary of Commerce addressed Cook Inlet crude oil development.

> The benefits that will ensue from these exports include increased incentives for investment in the exploration and development of domestic crude oil, transportation efficiencies, and material enhancements to the energy security of our allies. This initiative will also encourage other countries to remove trade barriers to U.S. goods and services. It does not affect our energy security as we retain the flexibility to react to changes in the world's available oil supply.[36]

BIS license policy for Cook Inlet crude oil exports in the *Code of Federal Regulations* states:

> Exports from Alaska's Cook Inlet. The licensing policy is to approve applications for exports of crude oil that was derived from the state-owned submerged lands of Alaska's Cook Inlet and has not been, or will not be, transported by a pipeline over a federal right-of-way subject to the [Mineral Leasing Act] or the Trans-Alaska Pipeline Authorization Act.37

The *Federal Register* notice states that BIS export licenses issued under these regulations will have a term no longer than one year and are subject to revocation "if there is serious interruption to available U.S. oil supplies."[38] According to the BIS, the agency approved a total of six licenses for the export of Cook Inlet oil in fiscal years 1986, 1987 (2), 1989 (2), and 1990.[39]

The state of Alaska receives royalties of approximately 12.5% of the oil and natural gas produced from its leases. These royalties may be taken as a share of the physical commodity—royalties "in-kind" (RIK)—or as a share of commodity value. According to the Alaska Department of Natural Resources, the state began exporting its RIK oil from the Cook Inlet to Taiwan in 1987. Exports to Taiwan continued through a series of one-year competitive auctions until 1991, when the last contracted deliveries were stopped under *force majeure* following the eruption of the Mount Redoubt volcano, which disrupted oil operations in the Cook Inlet area.[40] The total volume of RIK Cook Inlet crude oil exported to Taiwan between 1987 and 1991 was 3,587,088 barrels.[41] These exports never resumed. The exports by the state of Alaska correspond with the BIS licenses issued and appear to account for all oil exports from Cook Inlet. As *Figure 3* shows, while Cook Inlet exports may have temporarily increased crude production, they did not reverse the long-term decline of crude oil production in the region. Today, nearly all crude oil produced in Cook Inlet is supplied to a refinery near the city of Kenai, AK, which produces most of Alaska's gasoline as well as other fuels.[42]

OIL EXPORTS TO CANADA UNDER NAFTA

As discussed above, the United States-Canada Free-Trade Agreement Implementation Act of 1988 allowed up to 50,000 bbl/d of TAPS crude oil to be exported to Canada. These provisions were incorporated by reference in the North American Free Trade Agreement (NAFTA).[43] However, the regulatory record raises the question of whether the 1996 determination allowing unlimited exports from TAPS is in conflict with the earlier 1988 determination with regard to Canada. It could be argued that—as the superseding language—the 1996 determination does extend to Canada as the language is not country specific. Nonetheless, the 1988 determination capping the amount at 50,000 bbl/d is also still reflected in the BIS export regulations. Because Canada has apparently never imported U.S. crude oil under these specific provisions it is unclear whether the two seemingly different allowances represent an oversight in regulatory codification or the current licensing policy.

LIQUEFIED NATURAL GAS EXPORTS FROM KENAI LNG

The Kenai Liquefied Natural Gas Plant (LNG), located in Nikiski, AK, began operation in 1969 as the United States' first and, to date, only commercial LNG export facility. The plant was developed jointly by Phillips Petroleum (70%) and Marathon Oil (30%),[44] which supplied it with natural gas from their respective holdings in the North Cook Inlet gas field. In 1967 the Federal Power Commission[45] authorized the initial export of natural gas from Kenai LNG to Tokyo Electric and Tokyo Gas for a 15-year period (beginning in March 1969). This export authority has been extended a number of times since 1967, most recently by the DOE, which approved the export of 40 billion cubic feet of natural gas on April 14, 2014. Since its construction, Kenai LNG has exported over 2.2 trillion cubic feet of natural gas, nearly all of it to the two Japanese utilities.[46]

Although the Kenai LNG Plant continues to operate, ConocoPhillips announced plans to close it in 2011. At the time ConocoPhillips stated that "due to current market conditions ConocoPhillips has been unable to negotiate the commercial arrangements that would be necessary to allow for continued operation."[47] More specifically, the company cited its inability to secure long-term gas supplies from Cook Inlet (while facing declining local gas production and rising local demand) as the primary reason for Kenai LNG's closure.[48] The plant was closed over the winter in 2011; however it resumed shipments in May 2012 due to increased demand for Alaskan LNG in Japan after that country's earthquake, tsunami, and nuclear power plant closures (most notably in Fukushima). The plant suspended operations again in 2012 due to shortages of natural gas to meet local market demand in Alaska.[49] Kenai LNG's export authorization for these shipments expired in March 2013 and was not immediately renewed.

In April 2014, ConocoPhillips announced that it would reopen the Kenai LNG Plant because local gas utilities in Alaska had secured gas supply through the first quarter of 2018 and because the Cook Inlet area gas supply forecast had increased, providing a renewed opportunity to export Cook Inlet production in excess of local demand.[50] State officials had also asked the company to reopen the plant, reportedly to provide an incentive for gas producers to invest in continued Cook Inlet gas operations and to develop new discoveries. Kenai LNG resumed shipments in May 2014.[51]

As noted above, the DOE granted Kenai LNG export authority on April 14, 2014, including a blanket authorization to export LNG to non-free trade agreement (FTA) countries. The order authorizes approximately 40 billion cubic feet of cumulative exports for a two-year period.[52] In its application for the export authorization, ConocoPhillips argued that the natural gas to be exported was not needed to meet local demand and that exports would be a source of demand for Cook Inlet production during summer months when local demand is low. The company argued that this demand would help sustain the existing gas fields and provide an incentive for additional natural gas development in the region.[53] According to the DOE, no party to the authorization proceedings "submitted evidence to rebut the statutory presumption that the requested authorization is consistent with the public interest."[54]

ConocoPhillip's application for LNG export approval to non-FTA countries was filed on December 11, 2013. Ordinarily, the DOE would have placed this application in its administrative queue (Order of Precedence) behind nearly two dozen others pending before the agency, many for months or years.[55] However, the DOE expedited its consideration of additional exports from Kenai LNG at the request of ConocoPhillips. Reportedly, the agency viewed Kenai LNG as a fundamentally different project from the others due to its location, the relatively short term it was seeking (two years vs. twenty, for example), and the small volumes of gas involved compared to other projects in DOE's queue.[56]

ALASKA LNG PROJECT

Arctic Alaska has substantial natural gas resources. The U.S. Geological Survey (USGS) has estimated that conventional natural gas reserves on Alaska's North Slope potentially exceed 200 trillion cubic feet (Tcf), over eight times the total annual gas consumption of the United States.[57] The agency has also estimated up to 80 Tcf of undiscovered, technically recoverable onshore resources of shale gas on the North Slope.[58] Further, the USGS has estimated that the North Slope may contain up to 158 Tcf of technically recoverable natural gas in the form of methane hydrates.[59] Although methane hydrate resources cannot yet be developed because there are no commercially viable methods to do so, future technologies may make such production economically viable. Taken together, these vast natural gas resources—both proven and potential—on the North Slope have motivated the development of infrastructure projects to supply North Slope natural gas to southern Alaska and the lower-48 states via pipeline or to export markets in the form of LNG.

Ever since the first large natural gas reserves were discovered at Prudhoe Bay over forty years ago, Congress has been encouraging the development of these natural gas resources. The principal plan has been to construct a natural gas pipeline from the North Slope to the lower-48 states. Beginning with the Alaska Natural Gas Transportation Act of 1976 (ANGTA)[60] and continuing through the Alaska Natural Gas Pipeline Act of 2004, Congress has repeatedly affirmed a national need for an Alaska natural gas pipeline.[61] The presidential administrations of Jimmy Carter, Ronald Reagan, and George W. Bush also supported an Alaska gas pipeline project. President Obama voiced support for such a pipeline during his first month in office, when he described an Alaska gas pipeline as "a project of great potential ... as part of a comprehensive energy strategy."[62]

An Alaska gas pipeline from the North Slope to the lower-48 states would be, by some measures, the largest civilian construction project in the history of North America. When proposals were announced in the mid-2000s, developers estimated total project costs of $27 billion[63] to $30 billion.[64] At that time, lower-48 natural gas prices were at all-time highs, peaking above $10/MMBtu. However, with the recent growth of shale gas supplies in the lower-48 states, and the corresponding drop in North American natural gas prices below $4/MMBtu, an Alaska gas pipeline to the lower-48 states is not currently competitive. At the same time, prices for LNG in the international market have risen, for example, peaking above $16/MMBtu in Japan in 2012. These high international prices offer a potentially attractive export market for Alaskan LNG.

As an alternative to a lower-48 pipeline, Alaskan officials and North Slope producers have been pursuing a proposal to construct a pipeline from the North Slope to Nikiski, on the Kenai Peninsula, where a new LNG terminal would be constructed for LNG export. The project— Alaska LNG Project L.L.C. (Alaska LNG)—would consist of a gas treatment plant on the North Slope, an 800-mile pipeline, and a new LNG liquefaction plant and marine terminal at a total project cost estimated between $45 billion and $65 billion.[65] The pipeline would also provide several spurs along the route to supply natural gas to local Alaskan markets for heating, power generation, and other uses. Alaska LNG is being jointly developed by ExxonMobil, ConocoPhillips, and BP (the North Slope natural gas producers) as well as pipeline company TransCanada. The state of Alaska would also be an equity partner in the project under the provisions of a new state law enacted in April 2014 authorizing Alaska's participation in the project.[66]

On July 18, 2014, Alaska LNG's sponsors applied to the DOE for authorization to export approximately 929 billion cubic feet of LNG annually to both FTA and non-FTA countries over a 30-year term (beginning no later than 12 years after authorization is granted).[67] In submitting its application, Alaska LNG requested expedited review and exemption from new procedures under DOE's current Order of Precedence administrative queue on the grounds that the project's gas reserves were "more than sufficient" to serve local demand as well as exports and were "geographically isolated from the lower-48 states."[68] Energy Secretary Moniz reportedly agreed to "treat Alaska differently" from the other export applications because export of natural gas from Alaska would not affect markets in the lower-48 states and, therefore, the "public interest is not an issue" for the DOE.[69] Completion of an environmental review under the National Environmental Policy Act would still be required for final DOE authorization, as well as authorization from the Federal Energy Regulatory Commission (FERC) for the associated facilities. On September 5, 2014, Alaska LNG submitted a request to FERC to commence the commission's pre-filing process in preparation for filing an application for the facility authorization.

One complication for Alaska LNG is a requirement under ANGTA that

> before any Alaska natural gas in excess of 1,000 Mcf per day may be exported to any nation other than Canada or Mexico, the President must make and publish an express finding that such exports will not diminish the total quantity or quality nor increase the total price of energy available to the United States.[70]

In 1988, President Reagan issued such a finding, stating "the effects of exports of Alaska natural gas on American consumers would comply with the market criteria of [ANGTA] in the context of current and projected future energy markets and that such exports would be consistent with our comprehensive energy policy." [71] Although this finding was issued in the context of an earlier project (Yukon Pacific) to commercially develop North Slope natural gas, Alaska LNG's export application argues that it still applies to the current project and that a new finding under ANGTA is not required for Alaska LNG.[72] Whether President Reagan's finding applies to Alaska LNG is open to debate and will be an issue for consideration in DOE's review of the export application. Although Alaska LNG's developers have identified a proposed terminal site and pipeline route, this project is still largely conceptual with a number of unresolved engineering, technical, regulatory, environmental, economic, and permitting issues. Prior experience with the construction and operation of the Trans Alaska Pipeline System and the Kenai LNG terminal suggest that such a project would be technically feasible, although development time for the pipeline, in particular, could be long.[73] The North Slope is also a habitat for polar bears, whales, and other species listed under the Endangered Species Act; therefore further oil and gas development on Alaska's North Slope is a contentious political issue.

CONSIDERATIONS FOR CONGRESS

As stated in the introduction, to what extent the United States should allow overseas exports of crude oil and natural gas is the subject of ongoing

Congressional debate. Alaska's experience shows that—under a specific set of circumstances—such exports have been viewed as in the national interest both by Congress and successive presidential administrations.[74] Regional production trends, market prices, and local commodity demand have all been important considerations in establishing export policies for Alaska. It is also instructive to note that, even with long-standing export approvals in effect, exports of crude oil and natural gas from Alaska have been relatively modest. Ten years between crude oil export cargoes from TAPS, for example, suggests that production economics and competitive market forces will be the ultimate determinant of export volumes where exports are permitted.

In the context of broader U.S. oil and natural gas export policy the Alaska experience may raise several key questions. To what extent does the rationale for energy exports from Alaska—which is geographically isolated—apply to other U.S. supply regions? What is the interplay between overseas exports and the maritime shipping industry? What are the expectations for capital investment by developers to support export production and how might they affect the nation's overall oil and gas supplies? What are the environmental impacts of increased production for export? How might international trade agreements influence the U.S. government's ability to tailor its oil and natural gas export policies by region? As Congress continues its oversight of the nation's energy resources, understanding these issues may be important.

APPENDIX A. STATUTORY LIMITS ON TAPS OIL EXPORTS

Trans Alaska Pipeline Authorization Act of 1973 (P.L. 93-153 §101)

Limitations on Export

(u) Any domestically produced crude oil transported by pipeline over rights-of-way granted pursuant to section 28 of the Mineral Leasing Act of 1920, except such crude oil which is either exchanged in similar quantity for convenience or increased efficiency of transportation with persons or the government of an adjacent foreign state, or which is temporarily exported for convenience or increased efficiency of transportation across parts of an adjacent foreign state and reenters the United States, shall be subject to all of the limitations and licensing requirements of the Export Administration Act of 1969 (Act of December 30, 1969; 83 Stat. 841) and, in addition, before any crude oil subject to this section may be exported under the limitations and licensing requirements and penalty and enforcement provisions of the Export Administration Act of 1969 the President must make and publish an express finding that such exports will not diminish the total quantity or quality of petroleum available to the United States, and are in the national interest and are in accord with the provisions of the Export Administration Act of 1969: *Provided*, That the President shall submit reports to the Congress containing findings made under this section, and after the date of receipt of such report Congress shall have a period of sixty calendar days, thirty days of which Congress must have been in session, to consider whether exports under the terms of this section are in the national interest. If the Congress within this time period passes a concurrent resolution of disapproval stating disagreement with the President's finding concerning the national interest, further exports made pursuant to the aforementioned Presidential findings shall cease. (P.L. 93-153 § 101)

Export Administration Act of 1979 (P.L. 108-458 §7(d))

(d) DOMESTICALLY PRODUCED CRUDE OIL.—(1) Notwithstanding any other provision of this Act and notwithstanding subsection (u) of section 28 of the Mineral Leasing Act of 1920 (30 U.S.C. 185), no domestically produced crude oil transported by pipeline over right-of-way granted pursuant to section 203 of the Trans-Alaska Pipeline Authorization Act (43 U.S.C. 1652) (except any such crude oil which (A) is exported to an adjacent foreign country to be refined and consumed therein in exchange for the same quantity of crude oil being exported from that country to the United States; such exchange must result through convenience or increased efficiency of transportation in lower prices for consumers of petroleum products in the United States as described in paragraph (2)(A)(ii) of this subsection, (B) is temporarily exported for convenience or increased efficiency of transportation across parts of an adjacent foreign country and reenters the United States, or (C) is transported to Canada, to be consumed therein, in amounts not to exceed an annual average of 50,000 barrels per day, in addition to exports under subparagraphs (A) and (B), except that any ocean transportation of such oil shall be by vessels documented under section 12106 of title 46, United States Code) may be exported from the United States, or any of its territories and possessions, subject to paragraph (2) of this subsection.

(2) Crude oil subject to the prohibition contained in paragraph (1) may be exported only if—

(A) the President so recommends to the Congress after making and publishing express findings that exports of such crude oil, including exchanges—

(i) will not diminish the total quantity or quality of petroleum refined within, stored within, or legally committed to be transported to and sold within the United States;
(ii) will, within 3 months following the initiation of such exports or exchanges, result in (I) acquisition costs to the refiners which purchase the imported crude oil being lower than the acquisition costs such refiners would have to pay for the domestically produced oil in the absence of such an export or exchange, and (II) not less than 75 percent of such savings in costs being reflected in wholesale and retail prices of products refined from such imported crude oil;
(iii) will be made only pursuant to contracts which may be terminated if the crude oil suppliers of the United States are interrupted, threatened, or diminished;

(i) are clearly necessary to protect the national interest; and

(ii) are in accordance with the provisions of this Act; and

(B) the President includes such findings in his recommendation to the Congress and the Congress, within 60 days after receiving that recommendation, agrees to a joint resolution which approves such exports on the basis of those findings, and which is thereafter enacted into law.

(3) Notwithstanding any other provision of this section or any other provision of law, including subsection (u) of section 28 of the Mineral Leasing Act of 1920, the President may export oil to any country pursuant to a bilateral international oil supply agreement entered into by the United States with such nation before June 25, 1979, or to any country pursuant to the International Emergency Oil Sharing Plan of the International Energy Agency.

Trans-Alaska Pipeline Amendment Act of 1995 (P.L. 104-58 §202)

(f) Exports of Alaskan North Slope Oil—

(1) Subject to paragraphs (2) through (6), of this subsection and notwithstanding any other provision of law (including any regulation), any oil transported by pipeline over right-of-way granted pursuant to this section may be exported after October 31, 1995 unless the President finds that exportation

of this oil is not in the national interest. In evaluating whether the proposed exportation is in the national interest, the President—

(A) shall determine whether the proposed exportation would diminish the total quantity or quality of petroleum available to the United States;
(B) shall conduct and complete an appropriate environmental review of the proposed exportation, including consideration of appropriate measures to mitigate any potential adverse effect on the environment, within four months after the date of enactment of this subsection; and
(C) shall consider, after consultation with the Attorney General and Secretary of Commerce, whether anticompetitive activity by a person exporting crude oil under authority of this subsection is likely to cause sustained material crude oil supply shortages or sustained crude oil prices significantly above world market levels for independent refiners that would cause sustained material adverse employment effects in the United States.

The President shall make his national interest determination within five months after the date of enactment of this subsection or 30 days after completion of the environmental review, whichever is earlier. The President may make his determination subject to such terms and conditions (other than a volume limitation) as are necessary or appropriate to ensure that the exportation is consistent with the national interest.

(2) Except in the case of oil exported to a country pursuant to a bilateral international oil supply agreement entered into by the United States with the country before June 25, 1979, or to a country pursuant to the International Emergency Oil Sharing Plan of the International Energy Agency, any oil transported by pipeline over right-of-way granted pursuant to this section, shall, when exported, be transported by a vessel documented under the laws of the United States and owned by a citizen of the United States (as determined in accordance with section 2 of the Shipping Act, 1916 (46 U.S.C. App. 802)).

(3) Nothing in this subsection shall restrict the authority of the President under the Constitution, the International Emergency Economic Powers Act (50 U.S.C. 1701 et seq.), or the National Emergencies Act (50 U.S.C. 1601 et seq.) to prohibit exportation of the oil.

(4) The Secretary of Commerce shall issue any rules necessary for implementation, including any licensing requirements and conditions, of the President's national interest determination within 30 days of the date of such

determination by the President. The Secretary of Commerce shall consult with the Secretary of Energy in administering the provisions of this subsection.

(5) If the Secretary of Commerce finds that anticompetitive activity by a person exporting crude oil under authority of this subsection has caused sustained material crude oil supply shortages or sustained crude oil prices significantly above world market levels and further finds that these supply shortages or price increases have caused sustained material adverse employment effects in the United States, the Secretary of Commerce may recommend to the President who may take appropriate action against such person, which may include modification or revocation of the authorization to export crude oil.

(6) Administrative action with respect to an authorization under this subsection is not subject to sections 551 and 553 through 559 of title 5, United States Code.

APPENDIX B. STATUTORY LIMITS ON COOK INLET EXPORTS

Energy Policy and Conservation Act of 1975 (P.L. 94-163 §103)

DOMESTIC USE OF ENERGY SUPPLIES AND RELATED MATERIALS AND EQUIPMENT

SEC. 103. (a) The President may, by rule, under such terms and conditions as he determines to be appropriate and necessary to carry out the purposes of this Act, restrict exports of—

1) coal, petroleum products, natural gas, or petrochemical feedstocks, and
2) supplies of materials or equipment which he determines to be necessary (A) to maintain or further exploration, production, refining, or transportation of energy supplies, or (B) for the construction or maintenance of energy facilities within the United States.

(b) (1) The President shall exercise the authority provided for in Exemption, subsection (a) to promulgate a rule prohibiting the export of crude oil and natural gas produced in the United States, except that the President may, pursuant to paragraph (2), exempt from such prohibition such crude oil or natural gas exports which he determines to be consistent with the national interest and the purposes of this Act.

(2) Exemptions from any rule prohibiting crude oil or natural gas exports shall be included in such rule or provided for in an amendment thereto and may be based on the purpose for export, class of seller or purchaser, country of destination, or any other reasonable classification or basis as the President determines to be appropriate and consistent with the national interest and the purposes of this Act.

(b)In order to implement any rule promulgated under subsection (a) of this section, the President may request and, if so, the Secretary of Commerce shall, pursuant to the procedures established by the Export Administration Act of 1969 (but without regard to the phrase "and to reduce the serious inflationary impact of foreign demand" in section 3(2)(A) of such Act), impose such restrictions as specified in any rule under subsection (a) on exports of coal, petroleum products, natural gas, or petrochemical feedstocks, and such supplies of materials and equipment.

(c) Any finding by the President pursuant to subsection (a) or (b) and any action taken by the Secretary of Commerce pursuant thereto shall take into account the national interest as related to the need to leave uninterrupted or unimpaired—

(1) exchanges in similar quantity for convenience or increased efficiency of transportation with persons or the government of a foreign state,

(2) temporary exports for convenience or increased efficiency of transportation across parts of an adjacent foreign state which exports reenter the United States, and

(3) the historical trading relations of the United States with Canada and Mexico ...

End Notes

[1] Energy Information Administration, Short-Term Energy Outlook, September 9, 2014, Table 4a. Figures exclude Federal Gulf of Mexico production.

[2] Ibid, Table 5a. Figures exclude Federal Gulf of Mexico production.

[3] Department of Energy, "Summary of LNG Export Applications of the Lower 48 States," web page, September 19, 2014, http://energy.gov/fe/downloads/summary-lng-export-applications-lower-48-states.

[4] For further discussion see CRS Report R43231, Federal Permitting and Oversight of Export of Fossil Fuels, by Adam Vann, Daniel T. Shedd, and Brandon J. Murrill.

[5] See, for example: Charles Ebinger and Heather L. Greenley, "Changing Markets: Economic Opportunities from Lifting the U.S. Ban on Crude Oil Exports," Policy Brief 14-02,

Brookings Institution, September 2014; Harold Hamm, Chairman and Chief Executive Officer, Continental Resources, testimony before the Senate Committee on Energy and Natural Resources, Hearing on U.S. Crude Oil Exports: Opportunities and Challenges, January 30, 2014.

[6] In general, export of these products is not subject to the same restrictions as crude oil and natural gas.

[7] Alaska Department of Natural Resources, Cook Inlet Area Wide Oil and Gas Lease Sale: Final Finding of the Director, January 20, 2009, pp. 6-15.

[8] Ibid.

[9] P.L. 93-153, 43 U.S.C. §§1651 et seq.

[10] P.L. 108-458 §7(d).

[11] P.L. 100-449 §305(a).

[12] While an explanation of the precise figure of 50,000 bbl/d may be lost to history, it appears to have been related to the refinery capacity in Vancouver where it was presumed by Alaska officials the oil would be shipped.

[13] USCFTA, Annex 902.5(3)

[14] P.L. 100-449 §305(a).

[15] "Energy: Free Trade with Canada," Hearings before the Subcommittee on Energy and Power, Committee on Energy and Commerce, March 1, 1988, Serial No. 100-179. p. 4.

[16] President Ronald Reagan, "Presidential Findings Regarding the Export of Alaskan Crude Oil to Canada," December 31, 1988.

[17] Alaska Department of Natural Resources, personal communication, September 18, 2014.

[18] Very minor amounts also went through the Panama Canal to U.S. refineries on the Gulf of Mexico.

[19] See, for example: Samuel A. Van Vactor, "Time to End the Alaskan Oil Export Ban," Cato Policy Analysis No. 227, Cato Institute, May 18, 1995, http://www.cato.org/pubs/pas/pa-227.html.

[20] U.S. Congress, House Committee on Resources, Exports of Alaskan North Slope Oil, To Accompany H.R. 70, 104th Cong., 1st sess., June 15, 1995, H.Rept. 104-139 (Washington: GPO, 1995).

[21] Department of Energy, Exporting Alaskan North Slope Crude Oil—Benefits and Costs, June 1994, p. 1.

[22] Samuel A. Van Vactor, May 18, 1995. For further discussion see CRS Report R43653, Shipping U.S. Crude Oil by Water: Vessel Flag Requirements and Safety Issues, by John Frittelli.

[23] "Alaska Oil Export Ban Lifted," CQ Almanac, 1995, 51st ed., pp. 5-25-5-26, 1996.

[24] While the export ban was under debate in 1995, the United States was already exporting nearly 900,000 bbl/d of petroleum products—28% in the form of petroleum coke, which is used in making steel. Other exports were cross-border exchanges of refined products, as well as some crude, with Canada and Mexico. Trade in petroleum coke plus exports to Canada and Mexico accounted for 69% of all U.S. oil exports at the time.

[25] Trans-Alaska Pipeline Amendment Act of 1995, P.L. 104-58, 30 U.S.C. §185(s).

[26] Department of Commerce, Bureau of Export Administration, "Exports of Alaskan North Slope Crude Oil; Establishment of License Exception TAPS," 61 Federal Register 27255, May 31, 1996.

[27] President William J. Clinton, Memorandum to the Secretary of Commerce and the Secretary of Energy, April 28, 1996.

[28] Energy Information Administration, "How Much Oil Is Produced in Alaska and Where Does It Go?," web page, June 26, 2014, http://www.eia.gov/tools/faqs/faq.cfm?id=35&t=6
[29] Warren Cornwall, "Where Would ANWR Oil Go?" The Seattle Times, April 19, 2005.
[30] Jonathan Leff, "Tanker with Alaskan Crude Bound for South Korea as U.S. Seeks New Markets," Reuters, September 30, 2014.
[31] Energy Information Administration, "U.S Refinery and Blender Net Input of Crude Oil," web page, September 29, 2014, http://www.eia.gov/dnav/pet/hist/LeafHandler.ashx? n=PET&s =MCRRIUS1&f=A.
[32] Now called the Government Accountability Office.
[33] General Accounting Office, Alaskan North Slope Oil: Limited Effects of Lifting Export Ban on Oil and Shipping Industries and Consumers, GAO/RCED-99-191, July 1999.
[34] Resource Development Council, Inc., "Cook Inlet Oil," Resource Review, newsletter, Anchorage, AK, November 1985.
[35] Department of Commerce, International Trade Administration, "Exports of Crude Oil Derived from Alaska's Cook Inlet," 51 Federal Register 20252, June 4, 1986.
[36] Ibid.
[37] 15 C.F.R 754.2 (d).
[38] Department of Commerce, "Exports of Crude Oil Derived from Alaska's Cook Inlet," 50 Federal Register 52798, December 26, 1985.
[39] Bureau of Industry and Security, personal communication, September 8, 2014.
[40] Alaska Department of Natural Resources, Division of Oil and Gas 2009 Annual Report, May 2010, p. 36, http://dog.dnr.alaska.gov/Publications/Documents/AnnualReports/Section 2_2009.pdf.
[41] Ibid., Table II.8.
[42] Tim Bradner, "Cook Inlet Oil Production Increases 25% in Last Year," Alaska Journal of Commerce, August 28, 2014.
[43] NAFTA, Annex 608.2
[44] ConocoPhillips, a successor to Phillips Petroleum, bought out Marathon's 30% share in 2011 and is now the sole owner.
[45] The Federal Power Commission's regulatory authority over natural gas exports was transferred to the Secretary of Energy in 1977 by the Department of Energy Organization Act, 42 U.S.C. §§7151, 7172.
[46] Energy Information Administration, "Liquefied U.S. Natural Gas Exports to Japan, web page, August 29, 2014, http://www.eia.gov/dnav/ng/hist/n9133ja2a.htm; ConocoPhillips Alaska, "Kenai LNG Exports," web page, September 2, 2014, http://alaska.conocophillips. com/what-we-do/natural-gas/lng/Pages/kenai-lng-exports.aspx. In 2011, Kenai LNG shipped one cargo to China.
[47] ConocoPhillips Alaska, "Kenai LNG Plant," press release, February 2011.
[48] ConocoPhillips Alaska, "Additional Cargoes Extend Kenai LNG Facility Operations," press release, April 19, 2011.
[49] Yereth Rosen, "ConocoPhillips Restarts LNG Exports from Alaska," Reuters, June 14, 2012.
[50] ConocoPhillips Alaska, "ConocoPhillips Announces Resumption of Exports from the Kenai LNG Facility," press release, April 14, 2014.
[51] Tim Bradner, "State Asks ConocoPhillips to Reopen Idle Kenai LNG Plant," Alaska Journal of Commerce, September 12, 2013.
[52] Department of Energy, Office of Fossil Energy, Order Granting Blanket Authorization to Export Liquefied Natural Gas by Vessel from the Kenai LNG Facility near Kenai, Alaska to Non-Free Trade Agreement Nations, DOE/FE Order 3418, April 14, 2014, p. 1,

http://www.fossil.energy.gov/programs/gasregulation/authorizations/Orders_Issued_2014/ord3418.pdf.

[53] Ibid., pp. 9-10.

[54] Ibid., p. 18.

[55] See Department of Energy, Office of Fossil Energy, "Order of Precedence—Non-FTA LNG Export Applications," web page, March 24, 2014, http://energy.gov/fe/downloads/order-precedence-non-fta-lng-export-applications.

[56] Brian Scheid, "US DOE to Consider ConocoPhillips Alaska LNG Project Separately: Source," Platts, January 3, 2014.

[57] U.S. Geological Survey, Circum-Arctic Resource Appraisal: Estimates of Undiscovered Oil and Gas North of the Arctic Circle, Fact Sheet 2008-304, 2008, Table 1, http://pubs.usgs.gov/fs/2008/3049/fs2008-3049.pdf; U.S. Energy Information Administration, Annual Energy Outlook 2013, DOE/EIA-0383(2013), April 2011, p. 147.

[58] U.S. Geological Survey, Assessment of Potential Oil and Gas Resources in Source Rocks of the Alaska North Slope, 2012, Fact Sheet 2012–3013, February 2012, http://pubs.usgs.gov/fs/2012/3013/pdf/fs2012-3013_2-28-2012.pdf.

[59] U.S. Geological Survey, Gas Hydrate Resource Assessment: North Slope, Alaska, Fact Sheet, October 2008.

[60] 15 U.S.C. §719 et seq.

[61] For example, see P.L. 108-324 §103(b)(2)(A) "a public need exists to construct and operate the proposed Alaska natural gas transportation project."

[62] Erika Bolstad, "Obama Calls Alaska Gas Pipeline Promising," Anchorage Daily News, February 11, 2009.

[63] TransCanada Corp., Application for License: Alaska Gasline Inducement Act, November 30, 2007, p. 2.5-2. 2007 dollars.

[64] Wesley Loy, "BP, Conoco Join Forces to Pursue Gas Pipeline," Anchorage Daily News, April 9, 2008. 2008 dollars.

[65] ExxonMobil, "Alaska LNG Project Selects Lead Terminal Location," press release, October 7, 2013.

[66] Alaska Senate Bill 138, April 2014, http://www.legis.state.ak.us/PDF/28/Bills/SB0138G.PDF.

[67] Alaska LNG Project LLC, Application of Alaska LNG Project LLC for Long-Term Authorization to Export Liquefied Natural Gas, Department of Energy, Office of Fossil Energy Docket No. 14-96-LNG, July 18, 2014, p. 1.

[68] Ibid.

[69] Tim Bradner, "DOE to Streamline Export License for Alaska LNG Project," Alaska Journal of Commerce, August 19, 2014.

[70] 15 U.S. C. §719j.

[71] President Ronald Reagan, Presidential Finding Concerning Alaska Natural Gas, 53 Fed. Reg. 999, Jan. 15, 1988.

[72] Alaska LNG Project LLC, July 18, 2014, pp. 26-30.

[73] For details see Dan Sullivan, Alaska Dept. of Natural Resources, "Commercializing North Slope Gas," Presented to the Senate In-State Energy Committee, March 21, 2013, http://dnr.alaska.gov/commis/testimony/ Senate_Instate_Energy_3_21_13.pdf.

[74] Note that Alaska is not the only state where crude exports are allowed. For example, limited exports of heavy crude oil from California are also federally authorized. This and other exceptions to federal crude oil export restrictions are found at (15 C.F.R. §754.2(b)). They are outside the scope of this report.

In: U.S. Exports of Oil and Natural Gas
Editor: Frank G. Adams
ISBN: 978-1-63463-904-0

Chapter 4

FEDERAL PERMITTING AND OVERSIGHT OF EXPORT OF FOSSIL FUELS*

Adam Vann, Daniel T. Shedd and Brandon J. Murrill

SUMMARY

Recent technological developments have led to an increase in domestic production of natural gas and crude oil. As a result, there is interest among some parties in exporting liquefied natural gas (LNG) and crude oil to take advantage of international markets. This has placed new attention on the laws and regulations governing, and in many cases restricting, the export of fossil fuels.

In most cases, export of fossil fuels requires federal authorization of both the act of exporting the fuel and the facility that will be employed to export the fuel. For example, the export of natural gas is permitted by the Department of Energy's Office of Fossil Energy, while the construction and operation of the export facility must be authorized by the Federal Energy Regulatory Commission (FERC). Oil exports are restricted, but an export that falls under one of several exemptions can be authorized by the Department of Commerce's Bureau of Industry and Security.

Oil pipelines that cross international borders must be permitted by the State Department. Coal exports do not require special authorization

---

* This is an edited, reformatted and augmented version of a Congressional Research Service publication R43231, prepared for Members and Committees of Congress, dated August 4, 2014.

specific to the commodity; however, as with natural gas and crude oil, other generally applicable federal statutes and regulations may apply to the export of coal.

Restrictions on exports of fossil fuels could potentially have implications under international trade rules. They may possibly be inconsistent with the most favored nation requirement of Article I of the General Agreement on Tariffs and Trade 1994 (GATT 1994) if certain World Trade Organization (WTO) members are treated differently than others. Limits on exports could also potentially violate the prohibition on export restrictions contained in Article XI of the GATT 1994 if they prescribe vague and unspecified criteria for export licensing. However, an export licensing regime does not appear to constitute a "subsidy" to downstream users of fossil fuels under WTO rules.

Article XXI, the exception for essential security interests, may provide justification for potential violations of GATT Articles I and XI. The United States has traditionally considered this exception to be self-judging. However, it is possible that a panel or the Appellate Body might scrutinize the United States' use of the exception.

Article XX of the GATT provides additional exceptions that a member country may invoke if it is found to be in violation of any GATT obligations. For example, WTO members may maintain an otherwise GATT inconsistent measure if it is necessary to protect an exhaustible natural resource or necessary to protect human health or the environment. Article XIII potentially requires that if an otherwise GATT inconsistent measure is permitted to remain in force due to an Article XX exception, the measure must be administered in a nondiscriminatory manner. Export restrictions that treat WTO members differently would appear not to satisfy the potential nondiscriminatory requirements of Article XIII.

INTRODUCTION

Partly as a result of the increased use of horizontal drilling and hydraulic fracturing to extract natural gas from shale formations in the United States, the domestic supply of natural gas has increased relative to demand, leading to lower domestic prices. Domestic oil production has also increased after decades of decline thanks to new drilling technologies These production increases have generated new interest by some U.S. companies in exporting liquefied natural gas (LNG) to take advantage of relatively higher prices in world markets.[1] This new interest in exporting natural gas has also produced renewed interest in the laws and regulations governing the export of other fossil fuels, including crude oil and coal.[2]

This report reviews federal laws and the regulatory regime governing the export of natural gas, crude oil, and coal. This report provides an overview of federal laws and regulations and agency roles in authorizing and regulating the export of these fossil fuels. The report addresses several categories of federal laws and regulations, including (1) statutes that establish the authorization process for the actual export of any of the three listed fossil fuels; (2) statutes that govern the permitting of the facilities that export any of the listed fossil fuels; and (3) generally applicable trade statutes and treaties that affect exports of fossil fuels.

GENERALLY APPLICABLE EXPORT REQUIREMENTS

In general, transactions involving the export of items from the United States to a foreign country are subject to the Export Administration Regulations (EAR) enforced by the Department of Commerce's Bureau of Industry and Security (BIS).[3] However, transactions that fall within the scope of the EAR do not necessarily require an export license from BIS.[4] Whether an export license is required depends on several factors, including the nature of the item, its end use, and its ultimate destination.[5] The EAR provides instructions for exporters to follow when determining whether an export transaction is subject to the EAR and, if so, whether the transaction requires a license.[6]

Other general requirements may apply to transactions involving the export of items from the United States. For example, for exports of items subject to the EAR that do not take place electronically or in another intangible form, an exporter is required in certain circumstances to submit a Shipper's Export Declaration (SED) or Automated Export System (AES) Record to BIS and the International Trade Administration in the Department of Commerce's Bureau of the Census.[7] A declaration or record typically contains an identification of the exporter and the commodity being shipped; the date of exportation; and the country of ultimate destination, among other information.[8]

STATUTES GOVERNING AUTHORIZATION TO EXPORT FOSSIL FUELS

Crude Oil

The Energy Policy and Conservation Act of 1975[9] directed the President to prohibiting the export of crude oil and natural gas produced in the United States, except that the President may ... exempt from such prohibition such crude oil or natural gas exports which he determines to be consistent with the national interest and the purposes of this chapter."[10] The act further provides that the exemptions to the prohibition should be "based on the purpose for export, class of seller or purchaser, country of destination, or any other reasonable classification or basis as the President determines to be appropriate and consistent with the national interest and the purposes of this chapter."[11]

This general prohibition on crude oil exports and the exemptions to that prohibition are found in the BIS regulations on Short Supply Controls at 15 C.F.R. §754.2. The regulations provide that a license must be obtained for exports of crude oil, including those to Canada.[12] The regulations further provide that BIS will issue licenses for certain crude oil exports that fall under one of the listed exemptions, including (i) exports from Alaska's Cook Inlet; (ii) exports to Canada for consumption or use therein; (iii) exports in connection with refining or exchange of strategic petroleum reserve oil; (iv) exports of heavy California crude oil up to an average volume not to exceed 25,000 barrels per day; (v) exports that are consistent with certain international agreements; (vi) exports that are consistent with findings made by the President under certain statutes; and (vii) exports of foreign origin crude oil where, based on satisfactory written documentation, the exporter can demonstrate that the oil is not of U.S. origin and has not been commingled with oil of U.S. origin.[13]

The regulations also direct BIS to review applications to export crude oil that do not fall under one of these exemptions on a "case by case basis" and to approve such applications on a finding that the proposed export is "consistent with the national interest and the purposes of the Energy Policy and Conservation Act."[14] However, the regulations also seem to suggest that only certain specific exports will be authorized pursuant to this case-by-case review. The regulations provide that while BIS "will consider all applications for approval," generally BIS will approve only those applications that are either for temporary exports (e.g., a pipeline that crosses an international border before returning to the United States), or are for transactions (1) that result directly in importation of an equal or greater quantity and quality of crude oil; (2) that take place under contracts that can be terminated if petroleum supplies of the United States are threatened; and (3) for which the applicant can demonstrate that for compelling economic or technological reasons, the crude oil cannot reasonably be marketed in the United States.[15]

The regulations also provide for a few enumerated exceptions to the general license requirement.

These exceptions include foreign origin crude oil stored in the Strategic Petroleum Reserves,[16] small samples exported for analytic and testing purposes,[17] and exports of oil transported by pipeline over rights-of-way granted pursuant to Section 203 of the Trans-Alaska Pipeline Authorization Act.[18] These exports do not require a license from BIS.

Natural Gas/LNG

Section 3 of the Natural Gas Act (NGA) provides that "no person shall export any natural gas from the United States to a foreign country or import any natural gas from a foreign country without having first secured an order of the Commission authorizing it to do so."[19] This authorization is to be issued "unless, after opportunity for hearing, [the Commission] finds that the proposed exportation or importation will not be consistent with the public interest."[20] The Commission is further empowered to grant authorizations in part and to modify or place terms and conditions upon authorizations and to supplement its orders as appropriate.[21]

At the time of the NGA's enactment in 1938, the "Commission" referred to the Federal Power Commission. However, in 1977 the Federal Power Commission was dissolved and its responsibilities were transferred to the Department of Energy (DOE) as well as the Federal Energy Regulatory Commission (FERC), an independent agency operating within DOE, pursuant to the Department of Energy Organization Act.[22] Title III of this act transferred all functions of the Federal Power Commission to DOE except for those subsequently assigned to FERC in Title IV.[23]

Title III of the DOE Organization Act thus transferred the authority to authorize natural gas imports and exports from the Federal Power Commission to DOE. Title IV provides added clarity on this point. Section 402(f) of the act specifically states that "[n]o function ... which regulates the exports or imports of natural gas or electricity shall be within the jurisdiction of [FERC] unless the Secretary assigns such functions to [FERC]."[24]

Natural gas exporting responsibilities are handled by the Office of Fossil Energy within DOE.

The procedures for filing for authorization to import or export natural gas are set forth in DOE regulations found at 10 C.F.R. Part 590. The regulations establish filing requirements as well as the procedures for review of applications, including procedures that allow interested parties to participate in the process prior to the issuance of orders by DOE. The regulations also provide for an expedited filing and review process for one-time small volume imports and exports for "scientific, experimental or other non-utility gas use" without necessitating a permit.[25]

The Energy Policy Act of 1992[26] amended the NGA Section 3 generic requirement for a permit in order to export natural gas to create a more streamlined authorization process for imports from and exports to certain countries. Subsection (c) of Section 3 provides that the importation of natural gas from or exportation of natural gas to a country with which the United States has in effect "a free trade agreement requiring national treatment for trade in natural gas shall be deemed to be consistent with the public interest, and applications for such importation and exportation shall be granted without modification or delay."[27] This provision eased the authorization process for certain countries in the interest of free trade, including Canada and Mexico, the only countries with whom natural gas importation and exportation takes place via pipeline.

Section 3 of the NGA also protects the role of the states in the permitting decisions. State rights under various environmental statutes are protected with respect to both export authorization by DOE and permitting by FERC (discussed *infra*) in Section 3(d),[28] and Section 3(e) mandates the notification of relevant state authorities in order to gather their input during the process.[29]

Coal

Although the Energy Policy and Conservation Act of 1975 authorized the President to restrict coal exports,[30] the President does not appear to have exercised this authority to impose any significant export restrictions specific to coal. In fact, there have been legislative efforts aimed at expanding coal exports. For example, Section 1338 of the Energy Policy Act of 1992 directed the Secretary of Commerce to create a plan for expanding coal exports.[31]

Almost all U.S. coal exports pass through ports on the East Coast or in the Gulf of Mexico,[32] so laws and regulations applicable to such facilities would potentially affect coal exports. Such laws and regulations are briefly discussed below.

EXPORT FACILITY AUTHORIZATION

The previous section of this report discusses federal authorization of the export of natural resources, not the construction and operation of export facilities. However, in many cases approval for the export facility itself also must be obtained from the federal government. This section discusses various approval requirements for different types of facilities that enable the export of oil and natural gas.

Note that, in addition to the facility approvals described below, a facility used in the export or import of fossil fuels may require additional federal approvals or authorizations. For instance, construction and operation of ports in any navigable waters in the United States are regulated by the U.S. Army Corps of Engineers (ACE). In order to construct any port facility, permits must be obtained from ACE, which will review applications to see that they are in compliance with the Clean Water Act,[33] the Rivers and Harbors Act,[34] and the Marine Protection Research and Sanctuaries Act.[35] Because coal is generally not exported via a special facility designed to transport the commodity, there are no special facility permitting requirements applicable to coal exports, but facilities through which coal (or any fossil fuel) may be exported must satisfy these generic federal requirements.

Oil Pipeline Border Crossings

Crude oil can be exported either by pipeline or via tanker or other vessel. If an oil pipeline crosses the border with Canada or Mexico, the border crossing facility must be authorized by the federal government.[36] The executive branch exercises permitting authority over the construction and operation of "pipelines, conveyor belts, and similar facilities for the exportation or importation of petroleum, petroleum products" and other products pursuant to a series of executive orders. This authority has been vested in the U.S. State Department since the promulgation of Executive Order 11423 in 1968.[37] Executive Order 13337 amended this authority and the procedures associated with the review, but did not substantially alter the exercise of authority or the delegation to the Secretary of State in Executive Order 11423.[38]

Executive Order 11423 provides that, except with respect to cross-border permits for electric energy facilities, natural gas facilities, and submarine facilities:

> The Secretary of State is hereby designated and empowered to receive all applications for permits for the construction, connection, operation, or maintenance, at the borders of the United States, of: (i) pipelines, conveyor belts, and similar facilities for the exportation or importation of petroleum, petroleum products, coal, minerals, or other products to or from a foreign country; (ii) facilities for the exportation or importation of water or sewage to or from a foreign country; (iii) monorails, aerial cable cars, aerial tramways and similar facilities for the transportation of persons or things, or both, to or from a foreign country; and (iv) bridges, to the extent that congressional authorization is not required.[39]

Executive Order 13337 designates and empowers the Secretary of State to "receive all applications for Presidential Permits, as referred to in Executive Order 11423, as amended, for the construction, connection, operation, or maintenance, at the borders of the United States, of facilities for the exportation or importation of petroleum, petroleum products, coal, or other fuels to or from a foreign country."[40] Executive Order 13337 further provides that after consideration of the application and comments received:

> If the Secretary of State finds that issuance of a permit to the applicant would serve the national interest, the Secretary shall prepare a permit, in such form and with such terms and conditions as the national interest may in the Secretary's judgment require, and shall notify the officials required to be consulted ... that a permit be issued.[41]

Thus, the Secretary of State is directed by the order to authorize those border crossing facilities that the Secretary has determined would "serve the national interest."

Note that the source of the executive branch's permitting authority is not explicitly stated within the executive orders. Powers exercised by the executive branch are authorized by legislation or are inherent presidential powers based in the Constitution. Executive Order 11423 does not reference any statute or constitutional provision as the source of its authority, although it does state that "the proper conduct of foreign relations of the United States requires that executive permission be obtained for the construction and maintenance" of border crossing facilities.[42]

Executive Order 13337 refers only to the "Constitution and the Laws of the United States of America, including Section 301 of title 3, United States Code."[43] Section 301 of Title 3 provides that the President is empowered to delegate authority to the head of any department or agency of the executive branch. Courts that have addressed the legitimacy of this exercise of authority have found that it is a legitimate exercise of "the President's constitutional authority over foreign affairs and his authority as Commander in Chief."[44]

Natural Gas Pipeline Border Crossings

As discussed above, Executive Orders 11423 and 13337 explicitly exclude cross-border natural gas pipelines (among others) from their reach. Instead, permitting for these facilities is addressed in Executive Order 10485, which governs the issuance of Presidential Permits for natural gas facilities.[45] Executive Order 10485 designates and empowers the now-defunct Federal Power Commission:

1) To receive all applications for permits for the construction, operation, maintenance, or connection, at the borders of the United States, of facilities for the transmission of electric energy between the United States and a foreign country.
2) To receive all applications for permits for the construction, operation, maintenance, or connection, at the borders of the United States, of facilities for the exportation or importation of natural gas to or from a foreign country.
3) Upon finding the issuance of the permit to be consistent with the public interest, and, after obtaining the favorable recommendations of the Secretary of State and the Secretary of Defense thereon, to issue to the applicant, as appropriate, a permit for such construction, operation, maintenance, or connection. The Secretary of Energy shall have the power to attach to the issuance of the permit and to the exercise of the rights granted thereunder such conditions as the public interest may in its judgment require.[46]

In many ways, this authority resembles the authority over oil pipelines granted to the State Department in Executive Orders 11423 and 13337. However, as mentioned above, Executive Orders 11423 and 13337 do not describe the source of the executive branch permitting authority granted by the orders. Judicial opinions strongly suggest the permitting authority is an exercise of the President's "inherent constitutional authority to conduct foreign affairs."[47] By contrast, Executive Order 10485 cites federal statutes which may at least partially form the basis for the permitting authority granted to the DOE by the order. The order states that "section 202(e) of the Federal Power Act, as amended ... requires any person desiring to transmit any electric energy from the United States to a foreign country to obtain an order from the Federal Power Commission authorizing it to do so" and that "section 3 of the Natural Gas Act ... requires any person desiring to export any natural gas from the United States to a foreign country or to import any natural gas from a foreign country to the United States to obtain an order from the Federal Power Commission authorizing it to do so." These appeals to statutory authority should be considered and possibly addressed in any legislation seeking to amend the current Presidential Permit process for border crossings for energy facilities.

The Department of Energy Organization Act of 1977[48] eliminated the Federal Power Commission and transferred its functions to either the newly created DOE or the FERC, an independent regulatory agency within DOE. Section 402(f) of that act specifically reserved import/export permitting functions for DOE rather than FERC. As a result, DOE took over the FPC's Presidential Permit authority for border crossing facilities under Executive Order 10485 pursuant to the act. The authority to issue Presidential Permits for natural gas pipeline border crossings was subsequently transferred to FERC in 2006 via DOE Delegation Order No. 00-004.00A.[49]

LNG Export Terminals

Section 3(e) of the NGA, adopted in Section 311 of the Energy Policy Act of 2005,[50] assigns the "exclusive authority to approve or deny an application for the siting, expansion or operation of a Liquefied Natural Gas (LNG) terminal" to FERC.[51] Section 3 designates FERC as the "lead agency for the purposes of coordinating all applicable Federal authorizations" and for complying with federal environmental requirements.[52] Section 3(e) also directs FERC to promulgate regulations for pre-filing of LNG import terminal siting applications and directs FERC to consult with designated state agencies regarding safety in considering such applications.[53]

FERC implements its authority over onshore LNG terminals through the agency's regulations at 18 C.F.R. §153. These regulations detail the application process and requirements under Section 3 of the NGA. The process begins with a pre-filing, which must be submitted to FERC at least six months prior to the filing of a formal application. The pre-filing procedures and review processes are set forth at 18 C.F.R. §157.21. Once the pre-filing stage is completed, a formal application may be filed. FERC's formal application requirements include detailed site engineering and design information, evidence that a facility will safely receive or deliver LNG, and delineation of a facility's proposed location.[54] The regulations also require LNG facility builders to notify landowners who would be affected by the proposed facility.[55] To facilitate natural gas infrastructure projects, which includes LNG projects, FERC has adopted rules to provide "blanket certificates" that provide authorization to interstate pipelines to improve or upgrade existing facilities or construct certain new facilities pursuant to a streamlined process.[56]

Agreement on Tariffs and Trade

The Marrakesh Agreement Establishing the World Trade Organization (WTO) contains the agreements relating to international trade that are binding for all WTO members. Although there is no specific agreement relating to trade in energy products, such as liquefied natural gas, coal, or oil, the trade in these products is regulated under the General Agreement on Tariffs and Trade (GATT). Several of these sections could potentially impact a nation's ability to limit or restrict fossil fuels.

Article I—Most Favored Nation Treatment

Article I of the GATT 1994 requires that "any advantage, favour, privilege or immunity granted by any [WTO member] to any product originating in or destined for any other country shall be accorded immediately and unconditionally to the like product originating in or destined for the territories of all other [WTO members]."[57] Article I applies to all rules and formalities in connection with importation and exportation.[58] This broad category of rules and formalities appears likely to include prerequisites for exportation such as licensing requirements or other preliminary measures.[59] More favorable treatment given to imports from particular countries in the context of import licensing requirements has been held to confer an advantage within the meaning of Article I.[60]

Generally, this means that as soon as the United States provides for certain treatment of fossil fuel exports to one country, the United States has to treat exports to all other WTO members in the same fashion. A licensing regime that provided for more favorable treatment for exports of fossil fuels to some countries, but subjected other WTO countries to a slower process could potentially be inconsistent with Article I of the GATT.

However, there are exceptions to the Most Favored Nation Treatment requirements for Free Trade Agreements (FTA). Article XXIV of GATT 1994 allows countries to provide more favorable treatment to countries with which they have established an FTA.[61] In order to qualify for the Article XXIV exception, the FTA must meet certain requirements outlined in the Article. Most notably, the fee trade agreement must generally eliminate duties—such as tariffs—and restrictions on commerce between the parties to the agreement for "substantially all the trade in products originating in those territories."[62] Therefore, in order for an agreement to qualify, it is likely that the FTA would have to cover more than just energy products flowing between the two territories. However, if the countries have a qualifying FTA, more favorable treatment towards energy products moving between those countries could be included in that FTA without violating the GATT, although this may require a WTO panel to find that these provisions regarding energy products are essential to the agreement.[63]

Article XI—Export Restrictions

Article XI of the GATT covers import and export restrictions. Article XI:1 of the GATT bars the institution or maintenance of quantitative restrictions on exports to any WTO member's territory.[64] Quantitative restrictions limit the amount of a product that may be exported—common examples are embargoes, quotas, minimum export prices, and certain export licensing requirements. Under Article XI, duties, taxes, and other charges are the only GATT-consistent methods of restricting exports.[65] Any government action that expressly precludes the exportation of certain goods is inconsistent with the GATT.

Although there are few WTO panel decisions on export bans, panels have consistently found that import bans implemented through licensing systems violate Article XI.[66] This jurisprudence can be expected to inform any WTO panel decision on the GATT-consistency of export bans and licensing.[67] WTO Panel decisions have also held that "discretionary" or "non-automatic" licensing requirements are prohibited under Article XI—therefore, a licensing program that gives discretion to an agency to deny an export license to potential exporters on the basis of vague or unspecified criteria would violate Article XI.[68] Moreover, a GATT panel held that export licensing practices that cause delays in issuing licenses may be a restraint of exports that is inconsistent with Article XI.[69]

Articles VI, XVI, and the Agreement on Subsidies and Countervailing Measures—Export Restraints as Actionable Subsidies

A fossil fuel export licensing regime that restricts exports could have the effect of keeping domestic prices of fossil fuels lower than they otherwise would be. This raises the question of whether such a licensing program could be considered an actionable subsidy to downstream users of the fossil fuels such as members of the petrochemical industry. Under the GATT 1994 and the Agreement on Subsidies and Countervailing Measures (SCM Agreement), an actionable subsidy may be the subject of countervailing measures or challenge before a panel by a WTO member when the subsidy adversely affects the interests of that member.[70] Adverse effects might result if export restraints on fossil fuels lead to lower input costs for downstream manufacturers that use the fuels, giving the manufacturers' products a competitive edge over the products of the other members' manufacturers in domestic or foreign markets.

The SCM Agreement defines a "subsidy" as "a financial contribution by a government or any public body within the territory of a Member" that confers a benefit.[71] Under the agreement, one way that a "financial contribution" may occur is when a government directs a private body to sell goods to a domestic purchaser.[72] In U.S.—Measures Treating Export Restraints as Subsidies, the United States Trade Representative (USTR) argued before a WTO panel that a government's restriction on exports could be considered "functionally equivalent" to that government directing private parties to sell a good to domestic purchasers.[73] The USTR argued that this resulted in a subsidy to downstream producers that used the good as an input in their production processes.[74]

The panel rejected this argument, stating that although a restriction on exports of a good may result in lower prices for domestic users of that good, the restriction was not an explicit command or direction by the government to private parties to sell the good within the meaning of the SCM Agreement.75 This ruling suggests that future panels may be reluctant to find that a restriction on exports or a similar government intervention in a market is a "financial contribution" by a government. Thus, it seems unlikely that licensing procedures could constitute a subsidy under WTO rules, even if they lead to restrictions on exports.

Articles XX and XIII—General Exceptions

Article XX of the GATT provides for certain exceptions that a member country may invoke if it is found to be in violation of any GATT obligations. In order for the defense to be successful, the member country must show that its action fits under one of these general exceptions and that it satisfies Article XX's opening clauses, known as the "chapeau."[76] When dealing with trade in energy products, a country will most likely use the exceptions under Article XX(b) or XX(g). A country may justify a GATT inconsistent practice under Article XX(b) if the practice in question is "necessary to protect human, animal, or plant life or health."[77] Article XX(g) may permit otherwise GATT inconsistent measures that "relat[e] to the conservation of exhaustible natural resources if such measures are made effective in conjunction with restrictions on domestic production or consumption."[78] If a WTO member invokes an Article XX exception to the application of any quantitative export restrictions, Article XIII potentially requires that those export restrictions must be administered in a nondiscriminatory manner—that is, the restrictions must comport with the most-favored nation treatment discussed above.[79]

Article XXI—Security Exceptions

Restrictions on fossil fuels for reasons of international or domestic security that would otherwise violate the GATT 1994 may potentially be justified under the broadly worded exception for essential security interests contained in Article XXI.80 One paragraph of this article allows a member to take "any action which it considers necessary for the protection of its essential security interests ... taken in time of war or other emergency in international relations."[81]

While there is a lack of WTO case law on Article XXI, the nearly identical security exception under Article XXI of the General Agreement on Tariffs and Trade 1947 (GATT 1947) provides some guidance.[82] Under the GATT 1947, the contracting parties had broad discretion with respect to identifying an "emergency."[83] According to one source, each party was the judge of what was essential to its own security interests.84 Measures that parties sought to justify under Article XXI of the GATT 1947 included trade embargoes, import quotas, and suspensions of tariff concessions.85 Parties pointed to both potential and actual dangers as "emergencies" to justify measures otherwise inconsistent with the GATT 1947.[86]

With respect to who determines whether a WTO member's use of a national security exception is valid, the United States has taken the position that Article XXI is "self-judging."[87] That is, each member invoking Article XXI judges whether its use of the exception is valid. While there is currently no WTO case law on the use of Article XXI of the GATT 1994, some scholars have speculated that, in the future, a WTO panel or the Appellate Body may decline to defer to a WTO member's judgment that its use of Article XXI is appropriate and, instead, may subject a member's use of the exception to scrutiny.[88] For example, a panel may consider whether there is an emergency in international relations justifying national security screening for exports of fossil fuels to certain countries but not others.

NAFTA AND OTHER FREE TRADE AGREEMENTS

In addition to the GATT, the United States is party to numerous FTAs. It is beyond the scope of this report to discuss fully the provisions of each FTA signed by the United States. However, as an example, several FTAs require national treatment for trade in natural gas. These include FTAs with Australia, Bahrain, Canada, Chile, Colombia, Dominican Republic, El Salvador, Guatemala, Honduras, Jordan, Mexico, Morocco, Nicaragua, Oman, Panama, Peru, Republic of Korea, and Singapore.[89] The FTAs with Costa Rica and Israel do not require national treatment for trade in natural gas.[90]

As a further example, under the North American Free Trade Agreement (NAFTA) the United States has certain obligations related to energy trade with Mexico and Canada. Chapter 6 of NAFTA deals with "Energy and Basic

Petrochemicals." Chapter 6 reconfirms the Parties' obligations under the GATT and imposes additional obligations on the Parties, such as certain requirements for export taxes.[91] NAFTA also imposes barriers to invoking some of the general exceptions to the GATT.[92] For example, a country may only invoke the "conservation of exhaustible natural resources" exception if it does not result in a higher price for exports than for domestic consumption of the energy products.[93] These additional obligations illustrate that compliance with FTAs must be considered when establishing export regulations for energy products.

In addition, according to news reports, two proposed FTAs—the Trans-Pacific Partnership (TPP) and the Transatlantic Trade and Investment Partnership (TTIP)—could potentially include international obligations regarding the automatic approval by the United States of LNG exports to countries such as Japan (TPP) or the European Union (TTIP).[94]

LEGISLATION IN THE 113TH CONGRESS

S. 192, the Expedited LNG for American Allies Act of 2013, was introduced in the Senate on January 31, 2013, by Senator John Barrasso. An identical bill, H.R. 580, was introduced in theHouse of Representatives by Representative Michael Turner. The bills would amend Section 3 of the NGA to provide that expedited approval of LNG exports would be granted to four different categories of foreign countries: (1) nations for which there is in effect a free trade agreement (FTA) requiring national treatment for trade in natural gas; (2) a member country of the North Atlantic Treaty Organization (NATO); (3) Japan, so long as the Treaty of Mutual Cooperation and Security of January 19, 1960, between Japan and the United States remains in effect; and (4) "any other foreign country if the Secretary of State, in consultation with the Secretary of Defense, determines that exportation of natural gas to that foreign country would promote the national security interests of the United States."[95]

At least two other bills that pertain to the export of natural gas have been introduced. H.R. 1189, the American Natural Gas Security and Consumer Protection Act, was introduced in the House by then-Representative Ed Markey.[96] The bill would amend the NGA to require the Secretary of Energy to develop regulations for determining whether an export of natural gas from the United States to a foreign country is in the public interest for the purposes of issuing an export authorization.[97] Under the regulations, the public interest determination would have to be made after the Secretary's consideration of several factors, including the energy security of the United States; the ability of the United States to reduce greenhouse gas emissions; and an environmental impact statement issued under the National Environmental Policy Act that analyzes the impact of extraction of exported natural gas on the environment in communities where the gas is extracted.98 H.R. 1191, the Keep American Natural Gas Here Act, was also introduced in the House by then-Representative Ed Markey. Among other things, it would provide that the Secretary of the Interior could accept bids on new oil and gas leases of federal lands (including submerged lands) only from bidders certifying that all natural gas produced pursuant to such leases would be sold only in the United States.[99]

With regard to oil, H.R. 1190, the Keep America's Oil Here Act, was introduced in the House by then-Representative Ed Markey. The bill would provide that the Secretary of the Interior could accept bids on new oil and gas leases of federal lands (including submerged lands) only from bidders certifying that oil produced pursuant to such leases, and any refined petroleum products produced from that oil, would be sold only in the United States.[100] The bill would allow the President to waive this requirement for a lease in certain circumstances, including when a waiver is necessary under an international agreement.[101] In addition, S. 435, the American Oil for American Families Act of 2013, was introduced in the Senate by Senator Robert Menendez. The bill would ban the export of crude oil or refined petroleum products derived from federal lands (including land on the Outer Continental Shelf).[102]

Various other bills introduced in the 113th Congress seek to loosen restrictions on, or expedite the federal government's consideration of approvals for, the export of oil or natural gas to certain foreign countries.[103]

CONCLUSION

Recent advances in natural gas exploration and production technology have led to a newfound interest in the possibility of expanding U.S. fossil fuel exports. Such exports, and the facilities needed to conduct export operations, are subject to a panoply of federal laws and regulations.

These include the authorizations required by the Natural Gas Act, a generic ban on crude oil exports, and various laws and regulations applicable to construction and operation of export facilities. Currently, any party wishing to export fossil fuels must comply with these laws and regulations.

Under international trade rules, restrictions on exports of fossil fuels could potentially be difficult to reconcile with Articles I and XI of the GATT 1994. Article XXI, the exception for essential security interests, may be cited in order to justify potential violations of GATT Articles I and XI.

The United States has traditionally considered this exception to be self-judging. However, it is possible that a panel or the Appellate Body might scrutinize the United States' use of the exception.

Article XX of the GATT provides additional exceptions that a member country may invoke if it is found to be in violation of any GATT obligations. However, Article XIII potentially requires that if an otherwise GATT inconsistent measure is permitted to remain in force due to an Article XX exception, the measure must be administered in a nondiscriminatory manner. Export restrictions that treat WTO members differently would appear not to satisfy the potential nondiscriminatory requirements of Article XIII.

End Notes

[1] For more information about the potential for natural gas exports, see CRS Report R42074, U.S. Natural Gas Exports: New Opportunities, Uncertain Outcomes, by Michael Ratner et al.

[2] For more information about the potential for crude exports, see CRS Report R42465, U.S. Oil Imports and Exports, by Robert Pirog.

[3] Export transactions that fall within the exclusive jurisdiction of another federal agency are not subject to the EAR. 15 C.F.R. Part 734. In some cases, more than one federal agency may be responsible for exercising oversight over the export of a particular item.

[4] 15 C.F.R. Part 732.

[5] Id.

[6] Id.

[7] 15 C.F.R. Parts 30 and 758. The Bureau of Census uses the SED or AES to compile trade statistics. 15 C.F.R. §758.1. BIS uses the records for export control purposes. Id. Circumstances in which an SED or AES record is required to be submitted for the export of items subject to the EAR include when the items are destined for certain countries; when the export of the items requires submission of a license application under the EAR; and when the value of the exported commodities classified under a single Schedule B Number (or Harmonized Tariff Schedule number) exceeds $2,500. Id. Certain exceptions may apply. See id.

[8] U.S. Department of Commerce, A Basic Guide to Exporting 75 (1998).

[9] P.L. 94-163.

[10] 42 U.S.C. §6212(b)(1).

[11] Id. at §6212(b)(2). Note that the statute that provides the regulatory framework for promulgation and enforcement of these regulations, the Export Administration Act, expired in 2001. Since then the provisions of the act and the regulations have remained in effect via annual presidential orders issued pursuant to the International Emergency Economic Powers Act. For more information on this authority and other matters related to crude oil export, see

CRS Report R43442, U.S. Crude Oil Export Policy: Background and Considerations, by Phillip Brown et al.

[12] 15 C.F.R. §754.2(a).

[13] Id. at §754.2(b)(1).

[14] Id. at §754.2(b)(2).

[15] Id.

[16] Id. at §754.2(h).

[17] Id. at §754.2(i).

[18] 43 U.S.C. §1652; 15 C.F.R. §754.2(j).

[19] 15 U.S.C. §717b(a).

[20] Id.

[21] Id.

[22] P.L. 95-91.

[23] Id. at §301.

[24] 42 U.S.C. §7172(f).

[25] 10 C.F.R. §590.208.

[26] P.L. 102-486.

[27] 15 U.S.C. §717b(c).

[28] Id. at §717b(d).

[29] Id. at §717b(e)(2).

[30] 42 U.S.C. §6212(a).

[31] 42 U.S.C. §13367.

[32] Energy Information Admin, Quarterly Coal Exports October-December 2012 (March 2013), Table 13: U.S. Coal Exports by Customs District, available at http://www.eia.gov/coal/production/quarterly/pdf/t13p01p1.pdf.

[33] 33 U.S.C. §1344.

[34] 33 U.S.C. §403.

[35] 33 U.S.C. §§1401 et. seq.

[36] For tankers or other vessels conveying oil, the use of such facilities for exportation, in and of itself, does not require a permit akin to that required for oil pipelines that cross international borders. However, oil tankers or other such vessels must comply with other generally applicable export requirements, which are discussed elsewhere in this report.

[37] Exec. Order No. 11423, Providing for the performance of certain functions heretofore performed by the President with respect to certain facilities constructed and maintained on the borders of the United States, 33 Fed. Reg 11741. (August 20, 1968).

[38] Exec. Order No. 13337, Issuance of Permits With Respect to Certain Energy-Related Facilities and Land Transportation Crossings on the International Boundaries of the United States, 69 Fed. Reg. 25299 (May 5, 2004).

[39] Exec. Order No. 11423, 33 Fed. Reg. at 11741.

[40] Exec. Order No. 13337, 69 Fed. Reg. at 25299.

[41] Id. at 25230.

[42] 33 Fed. Reg. at 11741.

[43] 69 Fed. Reg. at 25299.

[44] Sierra Club v. Clinton, 689 F. Supp. 2d 1147, 1162 (D. Minn. 2010).

[45] Exec. Order No. 10485, Providing for the performance of certain functions heretofore performed by the President with respect to electric power and natural gas facilities located on the borders of the United States, 18 Fed. Reg. 5397 (September 3, 1953).

[46] Id.

[47] Sisseton-Wahpeton Oyate v. U.S. Department of State, 659 F. Supp. 2d 1071, 1081 (D.S.D. 2009).
[48] P.L. 95-91, 42 U.S.C. §4101 note.
[49] Available at http://www.ferc.gov/industries/electric/indus-act/siting/doe-delegation.pdf.
[50] P.L. 109-58.
[51] 15 U.S.C. §717b(e). Gas must be converted to LNG for export by means other than pipeline.
[52] Id.
[53] Id.
[54] 18 C.F.R. §153.8.
[55] 18 C.F.R. §157.6d.
[56] 18 C.F.R. §§157.201-157.218.
[57] General Agreement on Tariffs and Trade 1994, Art. I:1 (hereinafter GATT 1994).
[58] Id.
[59] See Panel Report, U.S.—Certain Measures Affecting Imports of Poultry from China, paras. 7.407, 7.410, WT/DS392/R (September 29, 2010) ("We conclude that 'in connection with importation' as used in Article I, not only encompasses measures which directly relate to the process of importation but could also include those measures ... which relate to other aspects of the importation of a product or have an impact on actual importation."). The same reasoning could apply to measures that have an impact on actual exportation, such as licensing requirements.
[60] Panel Report, EC—Regime for the Importation, Sale, and Distribution of Bananas, ¶ 7.193, WT/DS27/R/USA (May 22, 1997).
[61] GATT 1994, Art. XXIV.
[62] Id.
[63] See Appellate Body Report, Turkey—Restrictions on Imports of Textile and Clothing Products, ¶ 58, WT/DS34/AB/R (October 22, 1999).
[64] GATT 1994, Art. XI:1 ("No prohibitions or restrictions other than duties, taxes or other charges, whether made effective through quotas, import or export licences [sic] or other measures, shall be instituted or maintained by any contracting party on the ... exportation or sale for export of any product destined for the territory of any other contracting party.").
[65] Id.
[66] See Panel Report, Brazil—Measures Affecting Imports of Retreaded Tyres, WT/DS332/R (June 12, 2007) (holding a licensing system to be in violation of Article XI when a person would be ineligible to import tires based on where those tires came from); Panel Report, India Quantitative Restrictions on Imports of Agricultural, Textile and Industrial Products, WT/DS90/R (April 6, 1999).
[67] See Wen-Chen Shih, Energy Security, GATT/WTO, and Regional Agreements, 49 Nat. Res. J. 433, 451 (2009) (noting that it is likely that "the jurisprudence concerning quantitative restrictions on import in the interpretation and application of Article XI:1 also applies to exports.").
[68] See Panel Report, China—Measures Related to the Exportation of Various Raw Materials, WT/DS394/R (July 5, 2011) (holding that vague export licensing criteria allowed for too much discretion in granting licenses and that they were therefore in violation of Article XI).
[69] Panel Report, Japan—Trade in Semi-Conductors (May 4, 1988) GATT B.I.S.D. (35th Supp.), 31.
[70] GATT 1994, Arts. VI, XVI; SCM Agreement, Arts. V, VII, XI, XIX.
[71] SCM Agreement, Art. I.
[72] SCM Agreement, Art. I(a)(1)(iv).

[73] Panel Report, United States—Measures Treating Export Restraints as Subsidies, ¶ 8.22, WT/DS194/R (June 29, 2001).

[74] Id. at paras. 5.36, 5.48-.51.

[75] Id. at ¶ 8.42-.44.

[76] The "chapeau" requires, for example, that measures falling under these exceptions shall not be a disguised restriction on international trade. GATT 1994, Art. XX.

[77] GATT 1994, Art. XX(b). It is worth noting that the "necessary" requirement is a rather high standard to meet. Appellate Body Report, Brazil—Measures Affecting Imports of Retreaded Tyres, 150, WT/DS332/AB/R (December 3, 2007). In addition, a March 2014 WTO panel decision suggests that a WTO member may not rely on an Article XX(b) exception to justify restrictions on exports of "raw materials" when: (1) the member actually intends for the restrictions to benefit downstream domestic producers that use the materials as inputs; or (2) the member could implement increased pollution controls or other WTO-consistent alternatives to export restrictions. Panel Report, China—Measures Related to the Exportation of Rare Earths, Tungsten, and Molybdenum, paras. 7.169-7.171, 7.186-7.187, WT/DS431/R (March 26, 2014). This decision has been appealed to the Appellate Body.

[78] GATT 1994, Art. XX(g). Although "relating to" may be an easier standard to meet when relying on the exception, the measure in question must also operate in conjunction with domestic restrictions. Id. In addition, a March 2014 WTO panel decision indicates that a WTO member may not rely on an Article XX(g) exception to justify restrictions on exports of natural resources that have entered into commerce when doing so would interfere with international markets by allocating "quantities [of the resource] between foreign and domestic users." Panel Report, China—Measures Related to the Exportation of Rare Earths, Tungsten, and Molybdenum, ¶ 7.462, WT/DS431/R (March 26, 2014). This decision has been appealed to the Appellate Body.

[79] GATT 1994, Art. XIII.

[80] GATT 1994, Art. XXI.

[81] GATT 1994, Art. XXI(b)(iii).

[82] The GATT 1947 has been incorporated into the GATT 1994.

[83] See GATT Analytical Index – Guide to GATT Law and Practice 602-05 (6th ed. 1995) (hereinafter GATT Analytical Index).

[84] Id. at 600. See also Dapo Akande & Sope Williams, International Adjudication on National Security Issues: What Role for the WTO?, 43 Va. J. Int'l L. 365, 373 (2003).

[85] GATT Analytical Index at 602-05.

[86] Id. at 600.

[87] Akande & Williams, supra note 84, at 375-76.

[88] Id. at 383-84.

[89] Department of Energy, How to Obtain Authorization to Import and/or Export Natural Gas and LNG, http://energy.gov/fe/how-obtain-authorization-import-andor-export-natural-gas-and-lng.

[90] Id.

[91] North American Free Trade Agreement, Arts. 603, 604 (hereinafter NAFTA).

[92] NAFTA, Art. 605.

[93] Id.

[94] Mark Drajem & Edward Klump, Japan's Bid to Enter Trade Talks Opens Route For U.S. LNG, Bloomberg News, http://www.bloomberg.com/news/2013-03-17/japan-s-bid-to-enter-trade-talks-opens-route-for-u-s-lng.html; U.S. Trade Talks Could Deliver Cheaper Energy for

Europe, Reuters, (December 13, 2013), http://www.reuters.com/article/2013/12/16/eu-usa-trade-energy-idUSL6N0JP20L20131216.

[95] S. 192, §2.

[96] See also S. 2088 (Markey).

[97] H.R. 1189, §2.

[98] Id.

[99] H.R. 1191, §2.

[100] H.R. 1190, §3.

[101] Id. §4.

[102] S. 435, §2.

[103] E.g., H.R. 6; H.R. 4155 (Poe); H.R. 3760 (Poe); H.R. 4139 (Turner); H.R. 4349 (McCaul); S. 2083 (Udall); S. 2274 (Udall); S. 2494 (Udall); S. 2638 (Hoeven).

INDEX

D

E

F

G

H

I

J

P

Q

R

S

T

U

V

W

Y